Supplement to the Environmental, Health, and Safety Auditing Handbook

LEE HARRISON

Supplement to Environmental, Health, and Safety Auditing Handbook

Environmental Engineering Books

Aldrich
POLLUTION PREVENTION ECONOMICS: FINANCIAL IMPACTS ON
BUSINESS AND INDUSTRY
American Water Works Association
WATER QUALITY AND TREATMENT
Baker, Herson
BIOREMEDIATION
Cascio, Woodside, Mitchell
ISO 14000 GUIDE: THE NEW INTERNATIONAL ENVIRONMENTAL
MANAGEMENT STANDARDS
Chopey
ENVIRONMENTAL ENGINEERING IN THE PROCESS PLANT
Cookson
BIOREMEDIATION ENGINEERING: DESIGN AND APPLICATION
Corbitt
STANDARD HANDBOOK OF ENVIRONMENTAL ENGINEERING
Curran
ENVIRONMENTAL LIFECYCLE ASSESSMENT
Fiksel
DESIGN FOR ENVIRONMENTAL: CREATING ECO-EFFICIENT PRODUCTS
AND PROCESSES
Freeman
HAZARDOUS WASTE MINIMIZATION
Freeman
STANDARD HANDBOOK OF HAZARDOUS WASTE TREATMENT AND
DISPOSAL
Freeman
INDUSTRIAL POLLUTION PREVENTION HANDBOOK
Harrison
ENVIRONMENTAL, HEALTH, AND SAFETY AUDITING HANDBOOK,
SECOND EDITION
Hays, Gobbell, Genick
INDOOR AIR QUALITY: SOLUTIONS & STRATEGIES
Jain, Urban, Stacey, Balbach
ENVIRONMENTAL IMPACT ASSESSMENT
Kaletsky
OSHA INSPECTIONS: PREPARATION AND RESPONSE
Kolluru
ENVIRONMENTAL STRATEGIES HANDBOOK
Kolluru
RISK ASSESSMENT AND MANAGEMENT HANDBOOK FOR
ENVIRONMENTAL, HEALTH, AND SAFETY PROFESSIONALS
Kreith
HANDBOOK OF SOLID WASTE MANAGEMENT
Levin, Gealt
BIOTREATMENT OF INDUSTRIAL AND HAZARDOUS WASTE
Lund
THE MCGRAW-HILL RECYCLING HANDBOOK
Marriott
ENVIRONMENTAL IMPACT ASSESSMENT: A PRACTICAL GUIDE
Rossiter
WASTE MINIMIZATION THROUGH PROCESS DESIGN

Supplement to Environmental, Health, and Safety Auditing Handbook

Lee Harrison, Editor

President
Harrision Associates
Williamstown, Mass.

McGraw-Hill

New York   San Francisco   Washington, D.C.   Auckland   Bogotá
Caracas   Lisbon   London   Madrid   Mexico City   Milan
Montreal   New Delhi   San Juan   Singapore
Sydney   Tokyo   Toronto

Library of Congress Cataloging-in-Publication Data

Supplement to Environmental, health, and safety auditing handbook /
Lee Harrison, editor
    p.   cm.
  Includes index.
  ISBN 0-07-026921-1
    1. Industrial management—Environmental aspects—United States—
Handbooks, manuals, etc. 2. Environmental auditing—United States—
Handbooks, manuals, etc. I. Harrison, L. Lee. II. Environmental,
health, and safety auditing handbook.
  HD30.255.E58   1995 Suppl.
  363.7'06—dc21   96-47393
                                                                   CIP

McGraw-Hill

A Division of The McGraw·Hill Companies

1 2 3 4 5 6 7 8 9 0   BKP/BKP   9 0 1 0 9 8 7

ISBN 0-07-026921-1

*The sponsoring editor for this book was Zoe G. Foundotos, the editing supervisor
was Caroline R. Levine, and the production supervisor was Donald F. Schmidt.
It was set in Palatino by Victoria Khavkina of McGraw-Hill's Professional Book
Group composition unit.*

Printed and bound by Quebecor/Book Press.

Contents

Preface

As we approach the millennium, it is safe to say that environmental, health, and safety auditing has become standard operating procedure at most major corporations in North America and Europe. In fact, a growing number of companies with sophisticated, long-standing audit programs now view them, and their entire environmental management systems, as more than simply a means of providing insurance against calamitous events. They employ their environmental systems as tools for improving overall corporate performance in numerous areas.

Although environmental auditing began in the West and has been refined at the major industrial and consumer products manufacturing companies there, it is no longer seen as strictly the purview of companies in North America and Europe. Increasingly, these procedures are becoming accepted by companies and countries around the world, especially in Asia and Latin Amercia. The development of the ISO 14000 series of voluntary environmental standards, no doubt, is helping to drive this process. Indeed, companies seeking to do business worldwide should not be surprised to find ISO 14001 certification required even in countries where no elaborate environmental regulatory structure exists.

As one might expect—and as this supplement demonstrates—techniques employed by environmental auditors have become increasingly sophisticated over the last 15 years. Techniques have been developed or adapted to deal with both centralized and decentralized management structures, reengineered to evaluate entire environmental management systems, and expanded to measure sustainability in business to the extent of including an evaluation of the software used in EHS programs. Still, as with other areas of business, there is a need for continuous improvement, and it is with that thought in mind that we are publishing this supplement to the 1994 *Environmental, Health, and Safety Handbook.*

Lee Harrison

Contributors

Edward A. Blackford, AES Beaver Valley, Inc., Monaca, Pennsylvania (CHAP. 3)

Paul Burger, AES Placerita, Inc., Newhall, California (CHAP. 3)

Keith E. Kennedy, WMX Technologies, Inc., Oak Brook, Illinois (CHAP. 1)

Norman P. Moreau, Management Analysis Company, Washington, D.C. (CHAP. 6)

John S. Nagy, WMX Technologies, Inc., Oak Brook, Illinois (CHAP. 1)

Robert G. Newport, WMX Technologies, Inc., Oak Brook, Illinois (CHAP. 1)

Jane E. Obbagy, Vice President, Arthur D. Little, Inc., Cambridge, Massachusetts (CHAP. 5)

James L. Smith, Senior Environmental Management Systems Consultant (CHAP. 2)

David Wheeler, General Manager, Ethical Audit, The Body Shop International, Littlehampton, U.K. (CHAP. 4)

Supplement to Environmental, Health, and Safety Auditing Handbook

1

Environmental Audit Process Reengineering

From Compliance Verification to Management System Evaluation

John S. Nagy
Robert G. Newport
Keith E. Kennedy
WMX Technologies, Inc.
Oak Brook, Ill.

Introduction

Purpose

This chapter discusses the changes that occur to a corporate environmental audit program over time, changes that seek continuous improvement in the quality and efficiency of the program and that seek to ensure that the program is suited to the coverage, sophistication, and effectiveness of the environmental management programs and systems being implemented for the operations being audited. The chapter also will address concepts related to the evolution of audit programs and provide a detailed discussion of a case study—the continuing development of the environmental audit program being implemented by WMX Technologies, Inc.

Evolution of Corporate Environmental Audit Programs

Traditionally, environmental auditing within WMX and many other organizations has been directed toward detailed compliance verification—checking compliance with specific requirements. Over time, the WMX Environmental Audit Program has evolved from detailed compliance verification toward a management systems auditing approach. This evolution occurred in concert with the development of strong compliance management programs and systems within the company.

Figure 1-1 is a graphical representation of how corporate environmental audit programs must develop at the same rate that improvements are enacted to an organization's environmental compliance management programs and systems. In the early stages of development, the facilities audited have relatively less developed compliance programs, which could lead to conditions such as:

- Operating personnel are not aware of all requirements.

- Process and procedures to ensure compliance are not fully defined; compliance is not systematic, predictable, or reliable.

At this evolutionary stage, it is important for auditors to have complete lists of requirements and to check each requirement to make sure any instances of noncompliance are identified and corrected. It would not be uncommon during this stage for audit reports to include long lists of items that need to be addressed. In some cases the corrective actions that are implemented are short-term fixes, which may address the symptoms of a problem more so than the root cause of the problem.

As the sophistication and effectiveness of the compliance systems being implemented within an organization increase, the need for detailed compliance verification tends to decrease. For example, if a facility is aware of all requirements, has systems in place to prompt compliance-related activ-

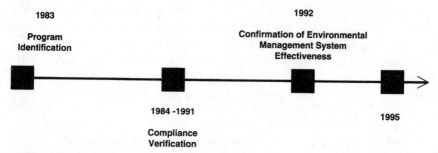

Figure 1-1. Environmental Audit Program evolution.

ities (e.g., a compliance calendar), does compliance self-checking, and has an effective training program, it is less likely that auditors can add value by reviewing specific requirements for which the facility has effective systems. Instead, the auditors can review the management systems, to make sure they are strong and complete, and focus detailed verification primarily in areas where the systems are relatively weaker or nonexistent. This leads to efficiencies in the audit process, and helps facilities to continually improve their processes and systems, based on the reviews conducted by the auditors.

Overview

This chapter describes the growth of the WMX Environmental Audit Program over time, showing strong correlation to the general model for the evolution of corporate audit programs. The reengineering of the WMX program involved a systematic process analysis initiative, followed by process redesign, testing, and implementation; each step is summarized. The chapter concludes with an assessment of the effects of the reengineering of the program and a forecast of the future direction the program will be taking.

A Case Study: The WMX Environmental Audit Program

Background

The WMX Environmental Audit Program is operated centrally from the corporate office, and covers several hundred operating locations. The objectives of the WMX Audit Program are to:

- Provide assurance to management that systems and controls are in place and being implemented at company facilities and operations to ensure continuing compliance.
- Assist facility managers in the identification of specific environmental compliance issues, and ensure that the issues which are identified are fully addressed and resolved.
- Evaluate compliance trends across the company's business groups.
- Work with the business groups to assess the need to strengthen company environmental policies and/or management systems.

Requirements reviewed during audits include federal and state statutes and regulations that address solid and hazardous waste management, air and water pollution control, and health, safety, and transportation requirements. The scope of the audit includes requirements established in permits, administrative rulings, contractual requirements, local ordinances, and company policies. Depending on the operations being audited, up to 25 subject areas are covered in audits, including such areas as construction, operations, air monitoring and emissions, generator standards, and surface water discharge management.

WMX environmental audits are conducted by audit teams of two to five auditors responsible for all preparation, on-site evaluation, and reporting activities. Time requirements for specific audits vary depending upon the type of operation being audited, the complexity of the regularity framework, and the size and composition of the team; most audits require 3–7 days per auditor for preparation activities, 3–5 days of work on site, and 2–5 days per auditor for reporting and "postaudit" work.

Program History

Prior to 1983, the company did not have an environmental audit program. In general, environmental management programs were in the early stages of development (i.e., staffing was minimal, the environmental management mission was narrowly defined, and formal compliance programs had not been universally implemented). Although company management stressed the importance of compliance, the means for achieving and consistently maintaining compliance had not been established; in addition, there were no consistent, objective compliance assessment mechanisms in the company. The prevailing assumption was that environmental compliance did not have to be managed any differently than other business elements.

WMX established an environmental audit program in 1983 in response to the identification of compliance management concerns within the company (see Fig. 1-2 for a graphical presentation of this stage and later developmental stages of the WMX Audit Program). Similar to the model for the evolution of compliance and auditing programs (Fig. 1-1), audits frequently found that compliance performance was not at the expected levels. The company recognized the need to strengthen the internal compliance focus and provide assistance to facility managers in their compliance assurance efforts. The environmental audit program that was established resulted in an improved compliance record, heightened compliance awareness throughout the company, management assurance that compliance was being (or would be) achieved, and effective measurement of the compliance status of individual facilities.

The audit process was oriented toward verification of compliance with all applicable environmental requirements. Audits focused on the

Figure 1-2. PACT cycle of continuous improvement.

present and the past, not on the future. Inspection activities were essentially a "snapshot" in time answering the question, "What was the condition of the facility on the days of the audit?" Document reviews tended to focus on the past, addressing issues such as, "Was the required report submitted?" and "Was the required monitoring conducted?" There was no coverage of health, safety, or transportation requirements. The emphasis was on the identification of compliance issues and tracking those issues through resolution. Root-cause analysis and management-system evaluations were not part of the process.

From the inception of the program until 1992, the basic element of the audit process remained unchanged (the program scope, however, was broadened periodically commensurate with regulatory changes and company growth). The process was heavily oriented toward compliance verification, which was consistent with the proficiency level of compliance management programs at the facility level. Although the mission and scope of audits remained relatively unchanged, quality assurance mechanisms were formally introduced into the audit process in 1988 after a self-evaluation identified the need to strengthen quality controls in several areas.

Another detailed self-evaluation of the audit process conducted in 1991 identified the need to advance the program from compliance verification auditing to a combination of management system evaluation and compliance verification. This advancement was related directly to the maturity and increased effectiveness of the company's compliance management programs. In addition, a mechanism for facilities to self-check compliance was introduced. Although audit reporting remained focused on specific facility-level compliance issues, audit teams began evaluating the compliance-related management systems employed at the operational level. Select health and safety requirements (principally right-to-know and emergency management) were also added to the scope of audits.

The third major audit process change, initiated in 1993, was developed through a structured process improvement initiative and implemented in early 1995. Process changes were prompted by the internal identification of improvement opportunities, with the goals of:

- Enhancing the quality and efficiency of the audit process, consistent with the total quality management principle of continuous improvement;

- Maintaining a leading-edge environmental audit program; and

- Recognizing the increasing effectiveness of environmental management systems and activities at company facilities (including a rigorous self-assessment program).

Audit process changes were targeted at achieving the following objectives:

- Reduction of the cost and cycle time of audits;

- Increased customer satisfaction; and

- Maximum use of audit resources in an expanding business environment.

The Process Improvement Initiative

WMX convened a process improvement team in late 1993 to evaluate audit program customer needs and environmental audit processes. The company's "continuous improvement roadmap" was implemented to ensure that the established objectives would be met. This roadmap is not unlike other total quality management and reengineering efforts; in fact, the most unique aspect of the roadmap was not the steps that were included but rather the top management support and tremendous enthusiasm for improving key business processes that accompanied this approach to process improvement. This support meant that unlike the process changes that occurred in 1988 and 1992, the Audit Department would be able to tap into additional resources and support for improving the audit process. The process improvement team interviewed process owners, auditors, and process customers, benchmarked with other organizations, and developed recommendations for the next evolution of the audit program. The revised processes were piloted in late 1994 and were rolled out in the first quarter of 1995.

Description of the Reengineered Environmental Audit Process

WMX's reengineered audit process has a more comprehensive scope, is more adaptable to changing requirements, and is better positioned to meet the organization's long-term needs. The key points associated with the major steps of the reengineered audit process are summarized below.

Audit Planning and Scheduling. Audits are conducted according to a "risk-based" schedule beginning in 1995. The risk-based approach means that audits are not scheduled according to a predetermined return frequency as had been the case since the inception of the program. Instead, several factors are considered in establishing the audit schedule, including:

- Complexity of the facility and operations
- Complexity of the regulatory environment
- Past compliance performance
- Continuity of the facility's management team
- Elapsed time since the previous audit

In addition, facilities are given less advance notice of upcoming audits than previously was the case.

Another important planning and scheduling change that emerged from the process improvement initiative is that most audit teams will include professional auditors and other compliance experts (e.g., environmental managers and engineers, and facility general managers who have strong compliance backgrounds) from throughout the company. The addition of these personnel to audit teams is expected to increase the "field experience" of audit teams, provide environmental auditing training to a broad spectrum of the company, and foster the exchange of best practices between facilities and business groups.

Audit Preparation. The steps involved and the time necessary for audit preparation have been significantly reduced, while the scope and quality of work have not been compromised. Rather than conducting independent updates of federal and state laws and regulations before every audit, the audit process will subscribe to an internal centralized service that will provide updated regulatory information throughout the company. A Quality Assessment/Quality Control (QA/QC) program has been established to ensure the accuracy and completeness of the regulatory information. This will allow auditors to devote more of their time to understanding the requirements that apply to a facility, and will ensure that the auditors and the facilities being audited are working from the same "rulebook."

On-Site Auditing. The scope of the on-site audit has been significantly increased and the focus has been shifted. Under the reengineered approach, the focus will shift to a more proactive assessment of compliance (i.e., a focus on management systems), followed by a detailed review of areas with a relatively higher risk of noncompliance. By approaching audits in this manner, the audit process will compile and evaluate information about the facility's management systems and the potential that the facility will remain in compliance in the future. At the same time, detailed compliance verification will not be lost; it will simply be focused in the areas that require such detailed review.

Reflecting this general approach of reviewing all management systems and conducting focused compliance verification, the reengineered on-site audit process consists of two phases:

- Phase I—An in-depth evaluation of the facility's compliance management systems and programs.

- Phase II—Detailed compliance verification focusing on areas where indicators of potential compliance concerns or management system weaknesses have been identified.

Phase I Management Systems Evaluations. Phase I audits apply a standardized approach for evaluating the management systems that facilities implement to assure compliance on a continuing basis. The procedures and criteria for management systems reviews closely reflect the work of Arthur D. Little, Inc. (*Environmental, Health, and Safety Auditor Handbook,* 1988). For the purposes of Phase I reviews, a management system is defined as a framework for guiding, measuring, and evaluating activities and performance. Management systems are a collection of programs, operations, people, documents, policies, guidelines, procedures, facilities, and equipment, designed to routinely or reliably achieve a desired objective (e.g., compliance with applicable requirements).

Phase I reviews include evaluations of management systems at two distinct levels:

- Facility-level management systems (e.g., training systems covering a broad span of subject areas and all facility employees).

- Environmental management systems related to specific programs or requirements. For example, for the management area "surface water discharge management," the management systems reviewed are:
  1. Identify point source discharges and ensure that all such discharges are permitted.
  2. Identify and ensure compliance with operational requirements for the control of discharges (e.g., operation of a treatment system or plant).
  3. Sample and monitor surface water discharges.
  4. Determine if discharges contain excessive quantities or concentrations of pollutants and follow up with appropriate actions if data seem to indicate potential problems.
  5. Report monitoring results and provide required notifications to federal, state, and/or local agencies, as required, and to maintain appropriate records.

The management systems reviewed as part of Phase I audits are evaluated in terms of their design and implementation. Following general industry practices for the evaluation of management systems, the following elements are examined:

1. Organization—The structure that establishes roles, responsibilities, accountability, and reporting relationships.

2. Guidance—The management of laws, regulations, permits, plans, policies, procedures, directives, and standards that provide direction regarding how activities are to be implemented.

3. Controls—The inspections, assessments, audits, reviews, and checks to determine conformance with requirements and standards.

4. Communication—The mechanism for collecting, managing, and reporting information within and outside of the operation.

Phase II Compliance Verification.   At the conclusion of Phase I work, the audit team determines which systems are strong and/or exhibited excellence or innovation and which systems are relatively weaker. Then Phase II detailed compliance verification is planned and conducted in the areas where potential system weaknesses were identified. Phase II audit procedures closely follow "traditional" compliance verification processes and include detailed checking against individual requirements. Management areas may also be selected for Phase II verification on a random basis, to selectively confirm the results of Phase I reviews where systems were found to be strong.

Reporting.   Reflecting the reengineered audit process, the key contents of audit reports are as follows:

■ Management system findings and recommendations (based on information collected and evaluated as part of Phase I reviews, and verified, as appropriate, through Phase II compliance verification);

■ Specific compliance issues and recommendations (identified through Phase II compliance verification activities); and

■ Audit score and opinion.

This information is intended to communicate to customers at facilities, group offices, and the corporate office, a sense of:

■ The current compliance status of the facility;

■ How effectively the existing management systems might assure compliance in the future; and

■ Areas that most warrant management attention.

The reengineered audit process allows the Audit Department to perform a more complete and incisive job of fulfilling these last two information needs. It is anticipated that evaluating and reporting on management systems will foster the continuous improvement in compliance programs, and a more proactive approach to identifying program areas that could be stronger, prior to an operation experiencing compliance problems. This impact of the audit program is similar in nature to how environmental auditing in the late 1980s helped push the organization toward greater overall levels of compliance awareness.

Since 1992, the Audit Department has used a numerical scoring system to measure or characterize environmental compliance performance. The scores are indicators of the percentage of audit areas reviewed that were generally in compliance. With the reengineered audit process, the department has maintained the scoring system, but it now complements the numerical scores with a more broad-based audit opinion. The scoring and opinion systems fulfill two key roles within the organization:

1. Facilitate the identification of compliance trends across the organization and over time.

2. Provide a performance-level gauge that can be used in internal compliance performance evaluations for specific parts of the organization and/or for business group or facility managers.

By moving the company's attention in the direction of the more broad-based audit opinion, which takes into account the effectiveness of compliance programs and management systems, the audit process will further encourage the development and implementation of strong programs and systems. Efforts are also being made within the company to identify additional measures of compliance performance, so there is less reliance specifically on audit scores in the evaluation of group or individual performance.

The Coexistence of Environmental Auditing and Environmental Compliance Management Programs

The WMX family of companies uses the acronym "PACT" to describe the components of an effective compliance management program. The components of PACT are:

- *Prevention.* The development and implementation of systems to identify applicable environmental requirements; communicating applicable requirements to appropriate operation personnel; and ensuring compliance with these requirements (i.e., systems to prevent compliance issues from occurring).

- *Assessment.* Mechanisms for continuously assessing the environmental compliance status of the operation and self-identifying issues that need to be addressed (i.e., inspections or self-audits).

- *Corrective actions.* Processes for ensuring that appropriate and timely corrective actions and actions to prevent recurrence are developed and implemented for all compliance issues.

- *Training.* The identification and implementation of regulatory and company training requirements, as well as site-specific training needs, and the provision of the necessary training.

The PACT model was fully defined and communicated throughout the company beginning in 1991, and by 1995 had been institutionalized within the organization. All operations that are covered under the WMX Environmental Audit program have developed and are implementing procedures and systems that cover the PACT components.

Key Compliance Management Tools

WMX has developed two compliance management tools that have been adopted for use at all WMX operations.[1] These tools are intended to

---

[1]The first tool is the Compliance Management System (CMS). This personal computer software system, developed by WMX, provides complete identification of applicable requirements in a form that is translated into tasks necessary for facility employees to complete to assure compliance. CMS tasks include a synopsis of the requirement, a regulatory and policy citation, the names of assigned personnel, the frequency with which the tasks must be performed, and other procedural information which helps to complete the tasks. The system prompts the completion of compliance-related tasks by established due dates or on preassigned intervals. Since the completion of tasks is also documented, the system effectively provides a self-check on compliance on a continuous basis.

assist facility, business group, and corporate managers in assuring compliance with applicable environmental, health, safety, and transportation requirements and managing corrective actions where compliance issues have been identified.

The second key compliance management tool is the "Compliance Action Reporting System" (CARS). This personal computer software system, which was also developed by WMX, facilitates the scheduling, tracking, resolution, and trend analysis of compliance issues across the company. CARS requires that a description of each identified compliance issue be entered into the system, along with planned corrective and preventive actions, a resolution "due date," and the person responsible. Each issue is then tracked in this system until final resolution. CARS information is accessible at the facility, business group, and corporate levels. Within the audit process, compliance issues identified during an audit are entered into CARS in a manner identical to issues identified through any other means (either internal or external). The system thereby affords a high level of assurance that audit-identified issues are given sufficient management attention and are tracked through final resolution.

PACT Evaluations

With the full implementation of PACT and use of the company's compliance management tools, company operations have established proactive compliance programs (i.e., preventing issues from occurring); most of the issues that still occur are self-identified through routine assessment mechanisms (e.g., inspections, self-audits, and CMS use). The audit scores generated as part of the audit process increased over the period 1992–1994, reflecting the successful implementation of PACT. It, therefore, became less important for the Audit Department to check each individual requirement and became more important to evaluate the management systems and compliance management programs being implemented at the operational level. Auditors confirm that PACT programs and systems are in place and evaluate their relative strength. The reviews are conducted by first building an understanding of the PACT systems being implemented and then comparing the design and implementation of these systems to established criteria.

PACT Evaluation Criteria

The following are highlights from the criteria applied to evaluate facility-level compliance management systems:

Prevention

- Personnel have clearly assigned responsibilities and understand their roles in assuring compliance.

- Requirements that affect facility operations have been identified and guidance has been developed to ensure compliance with these requirements.

- CMS has been fully implemented at the facility to effectively schedule, assign, track, and document the completion of compliance assurance tasks.

- The CMS database was complete and current.

- Other preventive systems have been established as necessary to ensure compliance (e.g., standard operating procedures).

- The facility's systems for preventing compliance issues have been reviewed periodically and modified to account for performance deficiencies, changes to requirements, and facility changes.

Assessment

- Personnel have clearly defined compliance assessment responsibilities.

- Mechanisms have been established to conduct periodic, consistent, objective, and documented assessments of the facility's compliance status.

- Required compliance assessment activities (e.g., self-audits and self-assessments, facility inspections, CMS assessments) have been conducted in a timely manner, and have effectively identified compliance issues that may exist at the division or facility.

- Compliance issues that have occurred have generally been self-identified, instead of being identified by a regulatory agency or other entity outside of the facility.

- Compliance issues have been communicated to appropriate levels of management.

Corrective Action

- Personnel understand identified compliance issues and have been assigned responsibility for resolution.

- CARS is used to schedule and document the resolution of all compliance issues and to communicate the resolution status to appropriate levels of management.

- Complete and effective corrective actions and actions to prevent recurrence have been developed based upon root cause analysis.

- Corrective actions and actions to prevent recurrence have been implemented according to established action plans and due dates.

- CARS data fields have been completed to accurately reflect the compliance issues; actions to be taken to resolve the issues; and the status of the issues, including both corrective and preventive actions. The information has been updated and communicated to appropriate levels of management on a monthly basis.

Training

- Personnel have been assigned training responsibilities and resources have been provided to assure training effectiveness.

- Regulatory, permit, and company-required compliance training has been provided to employees in accordance with applicable requirements; all training has been documented.

- The Division Compliance Coordinator (DCC) has been certified in accordance with the Group's DCC certification program and received additional training as necessary to complete assigned responsibilities.

- Mechanisms have been implemented to identify employee training needs and gauge the effectiveness of compliance-related training.

These criteria describe the expected or generally acceptable levels of performance. PACT systems reviewed by audit teams are rated against these criteria, and the results are a key factor in the determination of the overall audit opinion. A narrative on the results of the PACT reviews is included in audit reports on an exception basis; program elements that were substandard (i.e., below the criteria) or which demonstrated innovation or superior performance (i.e., noticeably above the criteria) are highlighted in reports.

Development of Compliance Management Program Standards

While the need for PACT programs has been communicated consistently since 1991, there has been an inconsistent understanding as to what level of implementation constitutes an acceptable PACT program. It is expected that the criteria being developed to review facility-level PACT systems soon will evolve into compliance management program standards. The criteria will be reformatted and revised as necessary, circulated for peer review within the company, and distributed as program guidance to the business groups and facilities. The standards-development process will also include a review of the WMX standards for conformity to external standards and expectations (e.g., the ISO 14000 standards and the U.S. Department of Justice Sentencing Guidelines). The standards will continue to be the basis for the review of facility-level management systems as part of WMX environmental audits, and will also be used by groups and facilities in planning and budgeting processes, and in conducting self-reviews.

The Compliance Infrastructure

Through root-cause analysis of compliance issues identified over the previous year, the Audit Department identified several characteristics or building blocks which, if not present, often resulted in relatively weaker environmental management systems and/or noncompliance. The department cataloged these building blocks and has characterized them as the "compliance infrastructure." They include such factors as commitment to compliance, personnel, resources, communication systems, and change management. As another component of the reengineered audit process, auditors now evaluate the extent to which a sound compliance infrastructure is in place at a facility. Similar to PACT reviews, a narrative on the results of compliance infrastructure reviews is included in audit reports on an exception basis—factors contributing to problems or which demonstrated excellence or innovation are described in reports. The compliance infrastructure is also a factor in the development of the audit opinion.

Compliance Management Best Practices

WMX has a relatively "flat" organizational structure with decentralized management and decision-making systems. The company has traditionally fostered an entrepreneurial business climate for the operating divisions. Most facilities have thrived in this climate, but the structure and practices have meant that communication across the organization is a continual challenge. A related outcome is that the company has many facilities, each setting up its own practices and systems, some of which are excellent and innovative and some of which are less strong.

As part of the reengineered audit process, it was recognized that the Audit Department can play a role in helping to identify best practices being implemented at facilities that have been audited, and communicating these practices to other parts of the organization. The department catalogs effective practices observed during audits and through the following means fosters the exchange of information on best practices:

- As new audits are conducted, if weaknesses are noted in facility-level management systems or in management systems related to specific environmental programs or requirements, information on effective practices observed at other facilities are included in audit reports as recommendations.

- The Audit Department distributes a newsletter, called the *Compliance Trendsetter*, to business group managers and to all operations included in the scope of the Audit program. Information on excellent or innovative management systems noted during audits is included in the newsletter.

The company periodically convenes a meeting of group compliance program managers. Best practices that have been cataloged in the previous 3 to 6 months are summarized and distributed in these forums.

It is expected that by fostering the exchange of information across the organization, the Audit Department can significantly increase the value added by its management system reviews.

Conclusions/Future Direction

Early results from the implementation of the reengineered audit process indicate that the cycle time for audit preparation has been significantly reduced. The time needed for audit work on site is highly dependent on the relative strength of the management systems being implemented at the facility. The time requirements are reduced for facilities with strong compliance-related management systems. Where systems are relatively weaker and more auditing time is needed for compliance verification, the duration of on-site auditing is approximately the same as under the previous process. Information on the levels of customer satisfaction with the reengineered process is not available to a degree that any significant conclusions can be made.

Near-term goals for implementing the reengineered audit process, and for coordinating the company's audit program and compliance programs, are as follows:

- Continue field testing/refinement of the revised audit protocol.

- Continue development of compliance program standards; coordinate with external standards development activities.

- Establish a formal training curriculum for compliance management systems.

- Analyze management system information collected during audits.

- Inventory best compliance management practices; redistribute information on best practices across the organization.

- Identify compliance management system trends possibly requiring additional attention.

- Continue to emphasize the importance of management system evaluations (such as root-cause analyses) in all internal compliance assessment activities.

The early results indicate that (1) the reengineered audit program is providing the benefits sought when the process improvement initiative was undertaken and (2) that the program is appropriately aligned with the sophistication and effectiveness of the company's environmental compliance programs. By "staying with the curve," as generally depicted in Fig. 1-1, confirming that strong programs and systems are in place at facilities, and providing services such as fostering the exchange of information on best practices, the WMX Audit Program can continue to add value to the organization even when overall compliance levels are generally high.

2

Auditing the Environmental Management System (ISO 14000)

James L. Smith

*Senior Environmental Management
Systems Consultant*

This chapter provides the reader with an overview of the following:

- ISO 14000 and environmental management systems
- The specifics of the ISO 14001 standard and how it relates to government regulations
- Auditing requirements and differences between ISO 14000 EMS audits and other types of audits
- ISO 14000 Auditor Qualification Requirements
- Future of ISO 14000
- ISO 14000 Resources

ISO 14000 Overview

Originally founded in 1946, the Geneva-based International Organization for Standardization (referred to as ISO after the Greek work meaning "equal") is comprised of 111 national standards writing bodies, such as the American National Standards Institute (ANSI) from the United States. As implied by its name, the ISO organization develops international standards for the purpose of equalizing the production of goods and services traded between nations.

Most ISO standards are product-specific. Like the American Society of Testing and Materials (ASTM) standards, they provide detailed technical specifications with respect to specific goods and services. However, in 1979, ISO formed the ISO Technical Committee (TC 176) to develop management system standards. These standards were originally related to quality assurance practices. The intent was to have one set of quality assurance standards that would be recognized by purchasers on a worldwide basis. In 1987, the first set of management system standards (ISO 9000) was issued.

Encouraged by the success of the ISO 9000 standards and in recognition of global concerns with respect to the environment, ISO formed a strategic action group (SAG) in 1991 to assess the need for a set of international environmental management systems (EMS) standards. At that time, several EMS initiatives were already under way by other organizations. These included the following:

- British Standards Institute standard BS 7750 *Environmental Management Systems,* issued in March 1992
- USA Standard ANSI/ASQC-E4, *Specifications and Guidelines for Quality Systems for Environmental Data Collection and Environmental Technology Programs,* issued in January 1995
- European Union's *Eco Audit* and *Eco-Labeling* Schemes
- Chemical Manufacturer's Association (CMA) *Responsible Care*

Program

Clearly, there was widespread interest in the early 1990s to develop standards for EMS programs. In the interests of promoting free trade and a level playing field between nations and industries, international standardization was needed. This is what the SAG team recommended.

Based on the SAG team's recommendations, TC 207 was formed in 1992. TC 207 met for the first time in June 1993 with 200 delegates representing worldwide interests for development of a single set of international EMS standards. The first set of ISO 14000 standards is due to be released in late-1996.

As with ISO 9000, ISO 14000 is not a single standard. Over the next few years, several ISO 14000 standards will be issued in the broad areas of EMS, environmental auditing, environmental labeling, environmental performance evaluation, life cycle analysis, terms and definitions, and environmental aspects in product standards. Currently, the standards closest to publication include the EMS standards and those for environmental auditing. This chapter will concentrate on those two areas.

Overview of the Environmental Management System (EMS) Requirements

Most of the interest in auditing ISO 14000 EMS systems will be with respect to ISO 14001, *Environmental Management Systems: Specifications with Guidance for Use.* This is because ISO 14001 is the only standard that TC 207 has developed as a *specifications* standard. The other standards in the ISO 14000 series are written as guidance documents.

The model used in development of ISO 14000 is similar to the Plan-Do-Check-Act (PDCA) model developed by practitioners in the quality field (Fig. 2-1). Specific paragraph numbers and subject areas from ISO 14001 are shown in Table 2-1.

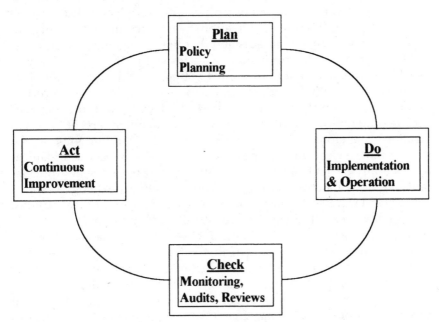

Figure 2-1. Plan, do, check, act cycle in relationship to ISO 14001.

Table 2-1. ISO 14001 Contents in Relation to Plan, Do,
Check, Act.

Plan	Environmental Policy (4.1)
	Environmental Aspects (4.2.1)
	Legal and Other Requirements (4.2.2)
	Objectives and Targets (4.2.3)
	Environmental Management Programs (4.2.4)
Do	Structure and Responsibility (4.3.1)
	Training, Awareness, and Competence (4.3.2)
	Communication (4.3.3)
	Documentation (4.3.4)
	Document Control (4.3.5)
	Operational Control (4.3.6)
	Emergency Preparedness and Response (4.3.7)
Check	Monitoring and Measurement (4.4.1)
	Nonconformance and Corrective/Preventive Action (4.4.2)
	Audits (4.4.4)
Act	Management Review (4.5) for Continuous Improvement

ISO 14001 can be used by companies to obtain third-party registration. It can also be used by an organization to self-declare itself as ISO 14000 compliant. It is intended to apply to all types and sizes of organizations. A key point is to understand that the management of the organization makes all determinations with respect to the applicability of ISO 14001. Management may use it to obtain certification of all of its facilities. On the other hand, management may decide to obtain certification of only one facility, or even part of a facility, and to either self-declare or to not comply at all with ISO 14000 in other parts of the organization. Therefore, in auditing ISO 14001, it is important that the auditor have a clear understanding of the scope to which the organization is applying ISO 14001. The auditor should not have any preconceived notions about what the organization ought to be doing. As stated in the standard, two organizations can be conducting similar activities and yet have very different, but acceptable, EMS programs.

In addition, a company that has an ISO 9000 quality management system may choose to use a part of that system in its EMS system. For example, certain activities, such as document control, are common to both standards. Since there is nothing unique about the manner in which documents are controlled between a quality management system and an environmental management system, it makes little sense to establish two separate systems. Therefore, it is expected that many companies that already have ISO 9000 quality systems will want to build on existing processes when developing an ISO 14000 EMS. This not only causes implementation to be smoother, but it also reduces the overall cost of establishing an ISO 14000 EMS.

Policy

The organization is required to have an environmental policy which is issued by top management. While the term *top management* is not defined by the standard, it should be realized that the standard is referring to management at a level that is recognized by the public, the stockholders, and the company's employees as having authority over the operations of the company. It cannot be a midlevel manager who is charted with carrying out the policies of the company. It needs to be at the level of CEO or president, and it should be supported by the company's board of directors.

This is not to say that the policy should be handed down "from on-high" with no input from those within the organization. Many companies use teams of employees to actually develop a policy with which the company can live. This is excellent; the more input to the policy, the better. A good, corporate citizen might even invite leaders of the local community to work with it in developing an effective, environmental policy. However, regardless of the number of people involved in developing the policy, it is essential that it actually be issued and fully supported by top management. Without that "buy-in" the EMS is a meaningless program.

The policy itself needs to make certain commitments as defined by the standard. These include the obvious commitments that a company will comply with all relevant environmental legislation and regulation affecting it. In other words, at the heart of the EMS, the company is agreeing with the local citizenry not to look the other way when the government's regulators are not around. It also includes a commitment toward continuous improvement in the EMS program as well as prevention of pollution. Prevention of pollution is an all-encompassing term that can refer to such measures as recycling, waste treatment, process changes, control mechanisms, efficient use of resources, and substitution of materials for a more environmentally friendly material.

For example, catalytic converters provide the most direct way of preventing unacceptable automobile emissions. However, there are many other ways of accomplishing this. They include using unleaded fuels (substitution of materials), computer-controlled fuel injection (control mechanisms), and design of more lightweight cars (efficient use of resources). Those are things that car manufacturers can do with their products to prevent pollution. City planners and highway engineers can also cut pollution by designing traffic patterns and public transportation systems that allow people to get to where they need to go with a minimum amount of driving. As individuals, we can make choices that also help to cut pollution. For example, we can choose to walk instead of drive or take mass transit or car pool to work. These latter methods may seem irrelevant to the car manufacturer; however, an enlightened car manufacturer could choose to work with city officials to find additional ways of reducing pollution. As a good, corporate citizen, the car manufacturer can have programs to encourage employees to carpool rather than drive to work separately.

None of those actions are *required* by ISO 14001, however. All ISO 14001 requires is that the company have a policy and that the policy address pollution prevention. The company is not required to pursue that requirement any further than what the law and regulations require. Therefore, it is again incumbent upon the auditor to first find out what is the company's policy. If the company says it has a program to encourage carpooling, then the auditor should review that to determine if it's being implemented and how well it's working. If the company does not make such a commitment, then it is *not* the job of the auditor to tell the company that it should.

ISO 14001 also requires that the policy be documented and communicated to all employees, as well as be available to the public. Some companies have devoted a wall in their reception area to the policy statement with the signatures of each employee around the policy statement. This provides excellent visibility to anyone entering the building of that company's environmental policy. It also shows that not only are the employees aware of the policy, but that they have bought into the policy. This is one way of communicating the policy, but it is not the only way. As an auditor, it is important to determine that employees understand and believe as true the environmental policy put forth by the company's management.

Planning

ISO 14001 introduces the term *environmental aspects* when discussing planning requirements. An environmental aspect is simply anything that the company does, makes, or sells that has some kind of interaction with the environment. This is obviously a very broad area. It can include the entire life cycle of a product from the initial mining or cutting of the natural resources used to the fuel consumed in shipping the product. In fact, for any given product, it is likely that no matter how well thought out the list of environmental aspects is, someone will always be able to find another one.

The standard requires identifying environmental aspects *that the company can control.* This does not necessarily mean, however, that the company need only consider its own processes and activities. For example, a company does not have to be in the business of timber cutting in order to make a decision about whether to use new wood or recycled pressed board in a product. In addition, while a company may not be able to do anything about a consumer who misuses or modifies a product after the point of sale, the company may be able to do something to inform the consumer of the potential damage to the environment through misuse or modification of a product. For example, if the company sells automobile oil to consumers, it can state on the oil container for the consumer not to pour unused or old oil down sewer drains because of the potential damage to local rivers, lakes, and wildlife. The oil company obviously cannot prevent someone from taking such an action, but it can inform those who may not be aware of the consequences of such action.

The purpose of identifying environmental aspects is to determine those significant environmental impacts that may impact the environment. The organization needs to identify those impacts and consider them in setting its environmental objectives. The standard also requires that this information be kept up to date. This means that when situations occur, such as design changes or changes in suppliers, the environmental aspects should be reassessed to determine if a new significant environmental impact might have been created.

Based on the significant environmental impacts which have been identified and on legal and other requirements to which the organization is obligated, environmental objectives and targets are required to be established. An *environmental objective* is an overall goal, which arises from the environmental policy, that the organization sets for itself. It should be quantified to the extent practical. An *environmental target* is a detailed performance requirement which is applicable to the organization, arises from the environmental objective, and needs to be set and met in order to achieve the objective. It too should be quantified where practical.

An example of this might be a company that sets a goal of contributing to improvement of the local environment. In order to accomplish this, one of its objectives could be to change the commuting habits of its employees. From this several targets might be established. Such targets could be:

1. One year from now, we will have increased by 50 percent the number of employees in the company van pool.

2. One year from now, we will have increased by 100 percent the number of employees taking public transportation to work.

3. We will increase the number of employees living within ten miles of the company through a combination of hiring practices and low interest loans offered to existing employees to live within ten miles of the company.

Note that the first two targets are quantified, and the success of the targets is measurable. The third target is not quantified. While the third target may be a worthwhile method of accomplishing the objective, it is a less desirable target if realistic expectations of its success cannot be assessed up front. For example, some employees may have chosen during the year to move closer to the company without the incentive of a low interest loan. If these are the only employees who take advantage of the low interest loan program, then the company will have wasted its money. Meanwhile, that same money could have been used somewhere else to more effectively improve the environment. But by establishing a target number of employees in advance, the company can assess what percentage of employees might move closer to the company during the year and set a target at a higher level. By doing so, the company can measure its progress during the year and determine if more effort needs to be made in providing incentives for such moves.

Once the objectives and targets are established, the company is required to establish and maintain programs for accomplishing those targets and goals. This includes assigning responsibilities throughout all applicable parts of the organization and establishing the means and time frames for achieving those objectives and targets.

To sum up the planning aspects of ISO 14001, the auditor should first study the environmental policy which the company has established. From the policy and the legal regulations to which the company is subjected, the auditor should be able to trace to the environmental aspects and impacts, the objectives and targets, and finally the programs established to address those objectives and targets. The company needs to have those planning steps clearly completed and documented before proceeding to the next step, which is the *doing* part of the PDCA cycle.

Implementation and Operation of the EMS

Although not as detailed as ISO 9000, much of the EMS's implementation and operation is similar to that of ISO 9000. It requires that roles and responsibilities be defined and documented, training needs be assessed and personnel trained accordingly, documents be controlled, controls be established over operational activities, and procedures be established to address the specific activities of the EMS. These controls are not new to auditors familiar with quality management systems. As stated earlier, if a company has an ISO 9000 quality management system, or similar type quality assurance program, then the ISO 14000 EMS can be integrated with that program. This saves costs and is easier for the employees of the company to understand and implement.

Some important considerations, however, need to be considered with respect to the EMS program. In auditing the implementation and operation of an EMS program, it's important to remember that the system being audited is implementing the environmental policy of the company. The auditor's obligation is to ensure that the policy, including its associated objectives and targets, is completely reflected in the company's operating and implementing procedures. At the same time, the auditor should not go beyond the policy to expect the company to be performing activities that are not within the scope of the company's policy statement.

Communication is an important part of a company's operations. Any company that undertakes the process of developing an EMS should make a concerted effort to communicate its actions, both internally and externally. Besides being a requirement of ISO 14001, good communications makes both the employees and the public aware of what the company is doing. Most people prefer to work for a company that is a good corporate citizen and mindful of the environment. Such employees are likely to be more dedicated employees.

While awareness training is important, ISO 14001 also discusses another type of communication. That is, procedures are required for internal communications between various levels and functions of the organization. These communications might include audit reports and results of EMS monitoring activities to responsible individuals. In addition, procedures are required for receiving, documenting, and responding to communications from external, interested parties. Such external parties might include a local citizen, a congresswoman, or a member of Greenpeace, for example. Requests might include a desire for information about the level of air pollution caused by the company's processes or a request for information about plans the company has to reduce future groundwater contamination. As a good, corporate citizen, the company should answer all such requests. In order to do so, the company should have orderly processes established for receiving and responding to any such requests. A company with an ISO 14000 EMS should not give the impression of "stonewalling" outsiders or being nonresponsive on environmental matters. Not only would this be contrary to the intent of ISO 14000, but it would lessen the public's overall perception of ISO 14000 environmental management systems.

While it should certainly be familiar to environmental, safety, and health professionals, emergency preparedness and response is an area of ISO 14001 that is not contained in ISO 9000. The ISO 14001 standard requires that the company

- Establish and maintain procedures that will identify the potential for accidents and emergency situations
- Establish methods for responding to such accidents and emergency situations
- Establish procedures for preventing and mitigating such accidents and emergency situations

It would be expected that the company would consider such impacts as accidental emissions to the atmosphere, and accidental discharges to local waterways or to the land, as may be applicable to that company's activities. Such impacts will normally arise from unusual occurrences during operations or from accidents and emergencies. Therefore, abnormal activities need to be considered in establishing emergency preparedness and response procedures.

Checking and Corrective Action

The next part of the PDCA cycle is the "checking" part, which is addressed by Section 4.4 of ISO 14001. A key part of the ISO 14001 requirements is the monitoring of those activities performed by the company that could have a significant impact on the environment. In order to properly audit this, the auditor needs to return to the environmental aspects and impacts that have been identified by the company and to review the established goals and targets which address those environmental impacts. From this information, the auditor can evaluate the scope and type of monitoring being performed to verify that it meets the objectives of the company's environmental policy. The auditor should also review the data being collected and verify calibration of the equipment being used to collect the data.

The remainder of Section 4.4 addresses the topics of nonconformance, corrective and preventive action, records maintenance, and EMS auditing. Specific audit requirements are discussed later. The other activities addressed by Section 4.4 should not be new to auditors familiar with management systems.

Management Review

ISO 14001 treats management review as a separate activity from that of checking and corrective action. In ISO 14001, the management review is the beginning of the continuous improvement process. It is required to be performed by top management; that is, the same management that established the initial environmental policy. The review is required to be documented, and required information and data must be collected and provided to top management to allow it to perform a proper review.

The management review is required to address possible changes to the policy, objectives, and other elements of the EMS as a result of audits and other changing circumstances. Because of the documentation requirements, the auditor should be able to actually verify the management review and *should not accept a statement from a company president that he holds weekly meetings with his vice presidents at which time the EMS is reviewed.* ISO 14001 clearly anticipates a more formal review.

Integrating ISO 14000 Requirements with Government Regulations

In the United States, many companies have a fear about establishing and implementing an EMS, particularly one that requires the company to audit its own processes and activities. This is due to the regulatory environment that has existed within the United States, whereby a company can be subject to fines and criminal penalties for known violations of laws and regulations. Therefore, the response of many companies has been to not audit, or to severely limit the scope of the audits to areas that are known to be fully acceptable.

The EPA, for its part, wants to encourage companies to perform both environmental management and environmental compliance audits. In order to so encourage companies, the EPA issued incentives for self-policing, through the Federal Register, on December 22, 1995. This policy is a refinement of its earlier March 1995 policy which offered powerful new incentives for regulated industries to discover, disclose, and correct violations of environmental law. Under the final policy, EPA will reduce civil penalties and not recommend criminal prosecution, in certain cases, of companies that voluntarily discover, disclose, and correct violations.

Based on this new policy, sometimes referred to as the EPA's auditing policy, companies should be more comfortable with establishing an EMS and performing audits of themselves. Therefore, ISO 14000 can be a valuable method for potentially reducing regulatory oversight and fines. It is recommended that companies who have had fears about self-policing activities obtain a copy of the new EPA auditing policy and review it to determine if they wouldn't be better served by developing an EMS program such as ISO 14000.

Audit Requirements

Reporting to the ISO/TC 207 Committee for Environmental Management are six subcommittees. One of these is SC2, Subcommittee on Environmental Auditing. Within SC2 are four working groups:

WG1.   Auditing Principles

WG2.   Auditing Procedures

WG3.   Auditor Qualifications

WG4.   Environmental Site Assessments

Based on the work of these working groups, SC2 proposes standards for publication as ISO 14000 standards. The initial work of the SC2 has focused on EMS auditing practices as opposed to environmental compliance auditing. Later standards can be expected to address environmental compliance.

There are currently three standards with respect to EMS auditing that will soon be issued. They are:

ISO/DIS 14010: *Guidelines for Environmental Auditing: General Principles.* This standard provides guidance regarding general principles of, and definition of terms used in, environmental auditing.

ISO/DIS 14011: *Guidelines for Environmental Auditing: Audit Procedures, Auditing of Environmental Management Systems.* This standard provides the procedures for actually conducting an environmental audit.

ISO/DIS 14012: *Guidelines for Environmental Auditing: Qualification Criteria for Environmental Auditors.* This standard provides guidance as to what constitutes a qualified environmental auditor, both internal and external to the company. It does not, however, address guidance for selection and composition of audit teams.

The term "DIS" refers to *Draft Internal Standard.* It has not yet been released as an approved document by the TC207. Care should be exercised in attempting to work to a draft standard.

The next two sections of this chapter address these standards in further detail. The next section discusses the key differences between ISO 14000 and other types of audits to which the auditor may be familiar. Following that section is a discussion of auditor qualification requirements.

Differences between ISO 14000 Audits and Other Audit Programs

Much of what is written in the three ISO 14000 draft standards on environmental auditing should be familiar to most experienced auditors. The principles of auditing remain the same between health and safety compliance, quality programs, environmental compliance, and EMS programs. However, the standard does introduce some changes that are specific to the standard and need to be understood by the ISO 14000 auditor.

Terminology

Terminology is the first area that the auditor new to ISO 14000 should study. For example, many auditors are familiar with, and use, the term *objective evidence*. However, this term does not appear in ISO 14000. Instead, there is the term *audit evidence*, which means much the same thing. ISO 14010 defines audit evidence as being verifiable information, records, or statements of fact. While the meaning between the two terminologies may be the same, the auditor needs to use the correct terminology for the ISO 14000 program. Not using the correct terminology could result in serious misunderstandings between the auditor and the audited organization.

Many auditors think of audit results in terms of writing audit findings that represent situations of nonconformance or noncompliance. In addition, many auditors have some freedom to include observations in their audit reports. This is not, however, the terminology that is used in ISO 14010. In ISO 14010, the term *audit conclusion* is introduced. This term refers to expressions by the auditor, which are made in the auditor's professional judgment or opinion, regarding the subject matter of the audit. Audit conclusions are based on and limited to reasoning the auditor has applied to the audit findings. An *audit finding* is the result of having evaluated audit evidence against the agreed-upon audit criteria. Audit findings provide the basis for the audit report. They are not necessarily noncompliances or nonconformances. However, they are based on verifiable audit evidence. The intent of the TC 207 SC2 in developing the terminology with respect to audit conclusions and audit findings is to limit the ability of the auditor to make recommendations which are not based on actual verifiable events observed during the audit process. Some auditors may not agree with this limitation, but it was considered necessary so that companies are not caught up in responding to auditor recommendations which do not have a valid basis.

In addition to terminology used in ISO 14010 and 14011, the auditor needs to understand environmental terminology and the terminology used by ISO 14001. For example, there was discussion earlier in this chapter about companies needing to identify their environmental aspects, and from that determine their significant environmental impacts. In order to properly audit a company to ISO 14000, the auditor must understand such terminology which is key to the overall EMS.

Requirements for an Environmental Audit

ISO 14010 is clear that the auditor does not operate in a vacuum. The auditor is expected to consult with the organization commissioning the audit in order to satisfy that client's objectives. If a consultant is hired to perform an internal audit of XYZ Corporation, that consultant should not develop a checklist and attempt to perform an audit of XYZ Corporation without working with XYZ Corporation's management to clearly define the exact objectives and scope of the audit. Auditors should not audit in any areas until they have been cleared with company management first.

The auditor also needs to assess the situation and make certain determinations before proceeding with the audit. These include the following:

- Is there sufficient and appropriate information available about the subject matter of the audit to make an audit worthwhile?
- Are there adequate resources to support the audit process?
- Is there adequate cooperation from the auditee?

If the answer to any of these questions is negative, then the auditor should not proceed with the audit.

Due Professional Care

Not all auditors may be familiar with the concept of *due professional care.* On the other hand, most auditors do in fact practice it. ISO 14010 devotes three paragraphs to the subject and states that it is the auditor's responsibility to exercise care, diligence, skill, and judgment when performing an audit.

There are many ways that this is carried out or not carried out during an audit. Most experienced auditors can cite numerous examples where due professional care was not exercised by someone. Some such situations could include:

Lack of care.   Auditors can easily leave company records lying around in conference rooms and other inappropriate places where unauthorized personnel may see them. Auditors have even been known to inadvertently take company records home in their briefcases and then fail to voluntarily return them.

Lack of judgment.   The auditor finds an activity that is not being performed in the normal way that the auditor is used to seeing that activity performed. Even though the activity is satisfactorily in accordance with all of the audit criteria, the auditor writes a finding of nonconformance because he does not take the time to fully understand other ways of performing the given activity.

Lack of skill.   Most auditors are skilled in auditing techniques; therefore, this is generally less of a technical problem than it is one of laziness. Properly, an auditor should select random samples to audit on her own. However, auditors will sometimes allow the auditee to pick the samples for audit. Another nonuse of an auditor's skills would be the auditor who chooses to audit training records by simply looking down a list of names as opposed to selecting names through a traceback method of first determining who is doing what.

These are some examples where due professional care may not be properly exercised by environmental auditors. ISO 14010 expects that auditors will maintain the highest professional standards at all times during the audit.

Reliability of Audit Findings and Conclusions

ISO 14010 actually addresses the topic of audit finding reliability. It states that audit evidence needs to be of sufficient quality and quantity that competent auditors, working independently of each other, would come to the same conclusions. This is difficult. It means that audits must be comprehensive enough to ensure that both identified problems really are problems and that real problems are not missed during the audit. Obviously, this cannot be possible without auditing 100 percent of the company's activities and processes. However, ISO 14000 expects that the error rate can be significantly reduced through a comprehensive audit.

ISO 14010 does not specify any statistical sampling requirements for determining audit finding reliability. And, in fact, statistical sampling may not be appropriate for EMS audits. However, some judgment is needed. If hiring a consultant to perform an audit and ten different consultants expect an audit to take a team of five auditors three weeks to perform, then the client organization would probably be violating this section of ISO 14010 to hire a consultant that will do the audit in one week with one auditor. Obviously, one auditor could not be expected to find in one week what five auditors would normally take three weeks to find.

The auditor is primarily responsible, however, for ensuring the reliability of audit findings and conclusions. The auditor needs to consider the limitations associated with the audit evidence and recognize the level of uncertainty that can arise from the audit findings and conclusions.

Audit Reporting

Most auditors forward their audit reports to the organization requesting the audit. ISO 14010 states that the auditor should also forward a copy of the report to the audited organization. This may be different in that most audit reports are submitted to the organization performing the audit which then may submit it to the audited organization.

Auditing Procedures

ISO 14011 provides a procedure for conducting the EMS audit. Because it is designed to be a procedure, no further procedure is required for the EMS auditor. It should be noted, however, that ISO 14001, while it requires audits to be performed, does not require that the ISO 14000 auditing standards be applied. Therefore, it is not necessary for an auditor to follow ISO 14011 as the procedure of choice. The organization requesting the audit may have its own separate procedure that it prefers the auditor follow. Or the auditor may work for a consulting organization which has a set of procedures for performing audits. If none of this exists, then the auditor and the requesting organization may want to choose ISO 14011 as the procedure to use. Note, however, that selection of the procedure to use is a joint decision between the auditor and the requesting organization; it is not the sole decision of the auditor.

There are several unique features to the ISO 14011 procedure of which auditors should be aware:

- The auditee is provided the opportunity to review the audit plan and to object to any provisions in the audit plan. Objections are required to be resolved to the satisfaction of all three parties (client, auditor, and auditee) before proceeding with the audit.

- During the collection of audit evidence, nonverifiable statements should be identified as such.

- Care is urged with respect to documenting audit findings of conformity so as to avoid any implication of total assurance of conformity.

The auditor and the requesting organization should thoroughly review ISO 14011 before deciding to make a commitment to it as their auditing procedure.

Auditor Qualifications

ISO 14012 establishes the guidelines for qualification of auditors and lead auditors. Again, it should be recognized that this is not a required standard, and an auditor can perform an ISO 14000 audit without these qualifications. For auditors interested in performing ISO 14000 EMS audits, however, meeting the recommendations of ISO 14012 provides a solid basis for starting. This will be especially true for internal audits, although the standard states that it can be applied to both internal and external audits. With respect to external audits, there are plans to develop third-party registration schemes, similar to ISO 9000, which may require other methods of qualification and certification.

Education, Experience, and Training

ISO 14012 recommends at least a secondary education with experience of four to five years or more in areas related to environmental science and technology; technical and environmental aspects of facility operations; relevant requirements of environmental laws, regulations, and related documents; environmental management systems; and audit procedures, processes, and techniques. It is clear from this listing that ISO 14012 is expecting the auditor to have education and experience in three key areas:

- Technical aspects of the areas being audited
- Laws and regulations affecting the areas being audited
- Auditing skills and techniques

One of these three areas is not really a substitute for the others. In order to be credible, the auditor needs a full understanding of all three areas. ISO 14012 identifies a mix of formal training and on-the-job training that the auditor should have in order to obtain the above knowledge in addition to the auditor's education and experience.

Personal Attributes and Skills

Auditors are expected to have certain personal attributes and skills. These include the following:

- Competence in clearly expressing concepts and ideas, both orally and in writing
- Diplomacy, tact, and an ability to listen
- Ability to maintain independence and objectivity
- Good personal organization
- Ability to reach sound judgments based on audit evidence
- Sensitivity to different cultures and conventions

The standard also states that the auditor should be able to communicate effectively in the language where the audit is being conducted. If not, and support of an interpreter is needed, the standard states that the interpreter should also be independent of the audited activities.

Lead Auditor

The standard allows for a couple of options in qualifying a lead auditor. One is that the individual can demonstrate necessary abilities as a lead auditor through interviews, observations, references, and/or assessments made under quality assurance programs. The demonstrations are made to the management of the audit program, and it is at their discretion that they determine an individual qualified to lead an audit or not.

The other method of qualification is more specific. It requires the lead auditor to first fully participate in the audit program for a total of 15 workdays of environmental auditing with three complete environmental audits. In addition, the individual would act as lead auditor, under the supervision and guidance of a qualified lead auditor, for at least one of the three audits. These criteria are expected to be met within a 3-year time frame.

Future of ISO 14000

The degree of acceptance of ISO 14000 on a worldwide basis, or even within a single country such as the United States, is not known at this time. However, there has been a worldwide movement to improve the environment. At the same time, governments are significantly cutting back on unnecessary expenses. This has placed more of the burden for environmental consciousness on private industry and citizens of the world. While there will always be some resistance, it appears that the overall movement of industry is toward improvement of the environment. This bodes well for standards such as ISO 14000 that help industry to focus on its environmental obligations and to establish management systems which will help them to manage their environmental programs.

The benefits of ISO 14000 are both societal and industrial. Society benefits when industry takes the initiative to "mind its own store." Society also benefits because industries in different nations are implementing the same programs to address environmental concerns. This helps to "level the playing field" with respect to trade.

Industry benefits as it learns that finding ways to reduce pollution can also decrease costs. On a more intangible level, industry builds up better relationships with its customers and employees. People become proud to work for, or otherwise be associated with, a company that is conscious about its impacts on the environment.

Because of these many benefits, it can be expected that ISO 14000 will be a major player in the world markets of tomorrow. It is a wise company that begins to look at ISO 14000 today and to figure out how to implement it into its business plans.

Additional Resources

The ISO 14000 standards are currently in draft form but will begin to be issued toward the end of 1996. They can be purchased from the American National Standards Institute (ANSI) at the following address and telephone number:

American National Standards Institute

Attn: Customer Service

11 West 42nd Street

New York, NY 10036

Telephone: 1-212-642-4900

In addition, the EPA's Auditing Policy (*Incentives for Self-Policing: Discovery, Disclosure, Correction and Prevention of Violations*) is published in the December 22, 1995 Federal Register.

Finally, there are numerous books, newsletters, and training courses provided by the private sector.

3

Environmental Auditing within a Decentralized Management Philosophy

Paul Burger

AES Placerita, Inc.
Newhall, California

Edward A. Blackford

AES Beaver Valley, Inc.
Monaca, Pennsylvania

Introduction

In the late spring of 1992 a reporting inconsistency was discovered within the data of the periodic reports mandated by permit at one of AES's domestic operating facilities.

The initial shock that something like this could happen within a company that prided itself on its environmental responsibility was supplanted by such concerns as "Is this a one time occurrence? An aberration?" to the more basic "What civil and potential criminal liabilities might exist for the company itself as well as those individuals who comprise it?".

The chief executive office of the AES Corporation in accordance with the board of directors moved quickly and decisively. A third-party audit performed by a national environmental auditing firm was to be performed under attorney-client privilege at all U.S. facilities within the next quarter. This action was followed by the creation of an environmental task force to provide an ongoing internal environmental auditing function.

Company Background[1]

The AES Corporation, formerly called Applied Energy Services, Inc., together with its subsidiaries, is engaged in the business of developing, owning, and operating independent (i.e., nonutility) electric power generation facilities. All of its current domestic plants are cogeneration facilities—a power generation technology that provides for the sequential generation of two or more useful forms of energy (e.g., steam and electricity) from a single primary fuel source such as coal or natural gas.

AES was formed in 1981 by Roger W. Sant and Dennis W. Bakke, chairman of the board and member of the chief executive office, and chief executive officer, respectively, and has grown to a company of approximately 1200 people. AES's primary objective is to help meet the need for electricity by being a clean, safe, and reliable power supplier. The company has pursued this objective by establishing a portfolio of plants that are largely solid-fuel fired, including both "scrubbed" pulverized coal and coal-fired circulating fluidized-bed cogeneration facilities. Through its subsidiaries AES owns and/or operates six facilities within the United States providing over 1000 megawatts of electricity and approximately 500,000 pounds per hour of process steam.

In 1991 AES became an international company through its participation in joint ventures which have purchased operating facilities in Northern Ireland and more recently in Argentina as a result of the privatization of the electric utilities within those respective countries. A similar joint venture has participated in the development of a grass roots project in England. Several projects are also in various stages of development and construction in India, Pakistan, and China.

Company Values and Practices

An important element of AES is its commitment to four major "shared" values: to act with integrity, to be fair, to have fun, and to be socially responsible.[1] These values are goals and aspirations to guide the efforts of the people of AES as they carry out the business purposes of the company.

In order to create a fun working environment for its people and implement its strategy of operational excellence, AES has adopted decentralized organizational principles and practices. As a result, approximately 7 years ago in 1989, AES reduced the number of supervisory layers in its organization to no more than two between the chief executive office and entry level positions everywhere in the company. While not deviating from this concept, a third supervisory level was added in the early 1990s as AES developed into a global power company with subsequent reorganization around a division concept. None of the company's plants have shift supervisors. Responsibilities for all major facility-specific business functions, including financing and capital expenditures, have been delegated to people at the plants. Criteria for hiring new AES people include a person's willingness to accept responsibility and AES's values as well as a person's experience and expertise. Every AES person has been encouraged to participate in strategic planning and new plant design for the company. The company has generally organized itself into multiskilled teams to develop projects, rather than forming "staff" groups (such as a human resources department, an office of legal affairs, or an engineering staff) to carry out specialized functions.

Environmental Task Force Development

Late in 1992, three people were chosen as co-chairpersons to lead a task force to organize and conduct environmental audits. Their initial efforts were to recruit additional task force members, establish objectives, provide relevant training to the task force, and decide upon both a scope and timetable for task force activities.

The first task of recruiting additional task force members was accomplished by means of a companywide posting for participation. No efforts were made to limit members to any specific educational background, job function, or geographic location. Response was overwhelming. While membership had to be limited to approximately 12 people to preserve the workability of the group, those members chosen represented all operating facilities (including international) as well as corporate headquarters. Backgrounds varied from operations and engineering to planning and finance. A representative from Arthur D. Little, Inc., the firm which had performed the attorney-client privilege audits, was also included as a full task force member to provide both guidance and credibility to task force functions.

Initial deliberations of the task force were directed toward establishing objectives which met the criteria that dictated its inception. Accordingly, a mission statement, included as Fig. 3-1, was developed establishing the objectives of this group. The first three objectives are addressed through the main activity of the task force: environmental auditing. The fourth objective has been approached through the initial formation of the task force as well as the practice of membership rotation over time. The fifth objective recognizes AES social responsibility

---

Environmental Task Force Objectives

1. To verify compliance with regard to federal, state, and local environmental laws and regulations and internal policies and procedures.

2. To keep the Environmental Committee of the Board of Directors and the Office of the CEO informed on the condition of each plant with respect to environmental issues.

3. To support all AES people in the management of environmental issues through the sharing of task force findings and recommendations, with the goal of strengthening each plant's environmental program and knowledge base.

4. To initiate and develop individual and group interest and awareness of the environment by dissemination of environmental knowledge, sharing ideas, and teaching each other.

5. To promote our vision of environmental excellence by striving for global leadership in providing clean and reliable generation to our customers.

6. To encourage networking, free thinking, and discussions of innovative solutions to environmental issues, while supporting the system of responsibility and accountability.

Figure 3-1. Environmental Task Force Objectives, October 1993.

and prevents the task force from limiting what it might accomplish. The sixth objective is in many ways a result of the task force in the sense that by meeting and evaluating current concerns it is in fact engaging networking and developing innovative solutions. The task force also recognizes that it is an advisory body and that responsibility and accountability remain at the level of each plant.

While task force members possessed varied job functions and diverse geographical plant locations, they did share one common thread. No one (except of course the Arthur D. Little, Inc. representative) had any practical environmental auditing experience. Accordingly, Arthur D. Little, Inc. was entrusted with the training function. While training attendance was mandatory for task force members, it was open companywide (space permitting) to any interested individual. This approach not only provided a more widely spread dissemination of knowledge but also created a pool of persons, who although not full fledged task force members, could serve on the various audit teams if needed.

It was ultimately decided that all six of AES's facilities operating in the United States would be audited during the first full year (1993) of the task force. With the training accomplished within the first quarter of 1993, this meant two audits in each of the subsequent quarters. This was a highly aggressive schedule in light of the fact that all task force members were serving in that capacity in addition to their normal activities at their respective plant locations.

Audit Scope

The initial task force audits made use of standard Arthur D. Little protocols covering the following functional areas, not all of which might be applicable to any given location:

- Air pollution control
- Solid and hazardous waste management
- Spill control and emergency planning
- Water pollution control
- Underground storage tank management
- PCB management
- Drinking water management

To the extent possible, the initial Arthur D. Little, Inc. audit would be used to establish a baseline from which to audit.

Audit Team

The audit teams have consisted of four people from AES locations other than the plant being audited. Each individual would cover one or two of the functional areas mentioned earlier.

One member of the team is designated as team leader. In addition to the usual functional area responsibility, the team leader managed all preaudit communications including the distribution of preaudit materials to the respective team members. The team leader periodically reviews team progress during the audit and has the additional task of writing the final audit report. For the audits performed in the first year task force co-chairpersons served as audit team leaders.

Additionally, an Arthur D. Little, Inc. representative also was a member of each audit team. While not participating in the audit itself, that person served in a consulting capacity. Not only were they able to assist in maintaining direction of the auditing function, they were also able to provide insight into past findings of the Arthur D. Little, Inc. audit which was not directly accessible to the other audit team members due to attorney-client privilege restraints covering that report.

In the initial auditing experience all team members could be regarded as beginners or rookies. However as the auditing year progressed and more experience was gained, a conscious effort was put forth in the latter audits to include one member on the team that had not participated in an audit to that point in time. In that way not only was the primary auditing function being realized but also the furthering of knowledge and training functions were being realized. The experience factor of the other AES team members allowed the Arthur D. Little representative additional time for mentoring the first time team member.

Audit Approach

The preaudit activities conducted by the team leader would begin with a questionnaire sent to the facility to be audited roughly 6 weeks prior to the scheduled audit. The team leader would then review the questionnaire following its completion and return. He or she would then request copies of appropriate permits, plan documents, plot plans, and process flow diagrams. Additionally, he or she would request a listing of key facility personnel along with their respective areas of responsibility. This information would then be shared with all appropriate team members.

Audits would typically be scheduled to begin on a Monday morning. Team members would arrive in the area the evening before, generally early enough to provide time for a brief meeting to discuss any last minute logistics or specific concerns. Following arrival on plant site Monday morning there would be an introductory meeting which would allow for introduction of team members as well as key plant personnel. The balance of the morning would consist of a plant tour. Monday afternoon as well as Tuesday and Wednesday normally would encompass the bulk of the audit consisting of document review, interviews with various plant personnel, representative sampling of monitoring data, and reporting. At the end of each of those days an informational meeting would be held to update plant personnel on the audit status including any potential findings. Typically these informal meetings are more detailed than any written formal report as those persons directly responsible are usually present and hence the meeting takes on more of an interactional focus and nature. Thursday morning and early afternoon finds the audit team wrapping up loose ends and preparing final summations of their respective functional areas. A wrapup meeting is then held which is a review of the entire audit. Audit team members would depart for their respective homes on Friday morning, although this schedule is generally kept soft in case there might be some unresolved issues that would carry over beyond Thursday.

After the Audit

During both the daily and final closing meetings all findings presented by the audit team are verbal only. Immediately after the closing meeting each AES audit team member would give a written accounting of any findings for his or her functional area to the team leader along with his or her working papers.

These written reports would be subdivided into the status of any findings from the prior Arthur D. Little audit followed by any new findings for a given functional area.

The team leader would then standardize and combine the individual findings of each respective audit team member into a composite final report. Although the report is essentially a report by exception detailing both regulatory and good management practice type findings, efforts are made to include positive observations in both the verbal presentations during the audit as well as in the final written report. Whenever a finding is presented a recommendation will be made by the audit team for its resolution. Following the recommendation, space is left to incorporate a formal plant response into this report.

Efforts are made to have the final written report drafted within 2–4 weeks after completion of the on-site portion of the audit. This draft is then circulated to the AES audit team members for their agreement with the presentation of findings. The draft is then reviewed by Arthur D. Little, Inc., as it relates to both presentation and accuracy of findings.

The report is redrafted as necessary and sent to the plant having been audited for its agreement of presentation as well as the generation of the plant responses to the audit team recommendations as noted earlier. Following this, the report is sent to legal counsel for their review after which the report is sent to the AES Office of the CEO. Only after their review is the report deemed final and subsequently sent to the AES Board of Directors. At that point any remaining copies of interim draft reports are destroyed and any remaining audit paperwork in the possession of audit team members is requested by the team leader for inclusion in the master file.

Task Force Activity

A separate function of the task force is to generate an annual environmental report for the AES Board of Directors. Within this report is a section containing annual environmental reports submitted by each respective plant. This provides an arena for updating audit follow-up as regards recommendations and plant responses which is a necessary segment of the audit cycle.

With all AES domestic plants having been audited in 1993, the schedule for 1994 included a second quarter audit of the two plants in Northern Ireland owned and operated under a joint venture. There was a much greater emphasis on the planning stage there as the goals and objectives of the AES partner (which happened to be a European entity) needed to be considered, as well as the major differences encountered from a regulatory standpoint. Accordingly the approach which had worked so well at the domestic plants had to be modified to reflect the somewhat varying conditions and goals that had to be met. Protocols were developed jointly with the AES partner specifically for these audits based on European Community Directives and the Statutory Rules of Northern Ireland. While the audit team did have to become familiar with somewhat varied legislation, they were at least spared the additional challenge of a foreign language. Team size remained constant at four members but with a slightly different makeup. AES and its partner own two distinct facilities in Northern Ireland, known collectively as NIGEN (Northern Ireland Generation). Accordingly each team was therefore comprised of a member of the AES Environmental Task Force who additionally served as team leader, a member from the AES partner, and two members from the NIGEN facility not being audited. While Arthur D. Little was utilized to review both the protocols prior to and draft report after the audit, there was no representative included on the actual audit team. Timing and structure of the audit itself was consistent with the domestic experience of the task force. The only difference in the reporting scheme was that the final report was presented to the NIGEN Board of Directors for subsequent release to the two parent companies.

For the balance of 1994 and 1995 the task force conducted one audit per quarter. No differentiation was made between domestic or foreign facilities as regarded that schedule. The rationale for choosing that frequency was twofold. First the individual time commitment to support that level of audit activity would not be overbearing when considered in the light of being additional to one's normal job function. Second, it would be often enough that auditing skills acquired would not be lost and there would be sufficient opportunity for task force members to acquire and polish those skills. With the current number of AES facilities worldwide, this frequency equated to about a 2-year audit cycle which is well within acceptable standards.

It had been decided that an annual training program would also be a favorable facet of an ongoing audit program. While the initial session was geared toward auditing skills and techniques, the 1994 session dealt more with regulatory issues with specific focus directed to the Clean Air Act which impacted all domestic AES facilities. Attention was also given to state laws impacting a particular plant within its jurisdiction. The main thrust in 1995 was the additional skills required of the position of audit team leader. At each training attendance was expected of all task force members and then open to anyone within AES on a space available basis.

Future Direction

As the task force gets into the third cycle of domestic audits (which in fact will be the fourth audit of a given facility considering the Arthur D. Little, Inc. audit done under attorney-client privilege), it is anticipated that the maturation of the program will tend to direct the audit more toward management and systems issues rather than those of pure compliance. This would be only natural in that recommendations to findings from the first two rounds of task force audits focused on establishing and implementing procedures, policies, etc., in order to address current issues.

In 1996 the task force schedule will reflect a stronger international influence. The NIGEN facilities will be visited for the second time and an AES facility in Argentina will be visited for the first time which introduces the added variable of language (Spanish) to be faced by the task force. Looking ahead to 1997, current plans call for a task force visit to mainland China to perform the audit function at jointly owned facilities there.

Extensive use of relatively generic Arthur D. Little, Inc. protocols was done as a matter of necessity in the first two rounds of domestic audits. It is anticipated that task force members will be able to refine those papers to more accurately reflect environmental performance for each AES plant by addressing appropriate state specific regulations. While facilitating future audits, this function of itself becomes a training tool. As already experienced with the initial NIGEN audits, development of specific protocols for each international audit can be viewed as the norm, thus placing a greater emphasis and resultant workload on the preaudit activities.

Rotation of task force members has become a serious issue with the task force entering its fourth year. As already noted, task force membership is in addition to one's regular duties within AES and thus can become burdensome over time or as one's other responsibilities may change. In order to ease this burden the schedule has been set so that each task force member can plan on performing one audit a year. While this solves the burden problem, it puts added pressure on training and maintaining proficiency at a function one participates in only once a year. International audits, with their inherent complexity, create the need to field a very experienced team to assure an accurate audit product. AES Environmental Task Force experience to date has been to field train its members in auditing and team leading skills at the domestic plants. This assures less surprise or uncertainty when subsequently graduating into the international arena. Additionally past members remain a resource by providing a pool from which to draw to either fill a schedule or strengthen a team as needed, while also allowing them to retain the auditing skills acquired while an active member.

Lessons Learned

While no member of the task force would claim to know all there is to know about environmental auditing with a little less than 3 years' experience, there have been a number of lessons learned. Some may tend to be of a "common sense," nature while others are more specific to AES and its culture.

- The four-person team (AES members) seems sufficient for a Monday morning start and Thursday afternoon wrapup, without the need of excessively long work days in between.

- Schedules should be coordinated as much as possible at the plant-to-be-audited's convenience. The audit team is in fact disrupting normal activity, so it is important to keep that disruption to a minimum in order to maintain a high level of cooperation during the audit.

- Whenever possible, task force meetings, training, etc., are held at plant locations. It is important within AES's culture that staff functions do not exist. This action supports the attitude that the task force is a function of the plants collectively rather than any corporate body.

- The spirit of "openness" on the part of plant people realized thus far indicates acceptance of the task force mission as well as a desire to learn and improve.

- The use of technicians as auditors has been very positive in both the aspects of performance as well as personal development.

- Daily lunchtime updates on plant site inviting a wide cross section of plant personnel provides a good arena for information exchange as well as increasing environmental awareness.

- The Arthur D. Little representatives have been impressed by the objectivity of the audit teams in maintaining "arms-length findings."

- The presence of the Arthur D. Little people has been helpful from a "coaching" standpoint but not overbearing from a "we need to do it this way" in order to be able to accredit the final report.

- There is enough difference between AES various plant sites that it is important that interplant competitions do not develop based on audit results as that runs counter to both the task force mission and the culture of AES. Conversely there are enough similarities that the networking and sharing of ideas is essential for both the overall attainment of the task force goals and satisfying the AES ideal of being the best it can be.

- When doing an international audit, the auditor needs to be able to accept the mindset of a national from that country to avoid the pitfall of auditing against U.S. standards.

- There is no set format when it comes to an international audit. The task force needs to be responsive to the specific needs and goals for that particular audit and react appropriately to ensure that a true and accurate product is generated.

Conclusions

The accomplishments of AES's Environmental Task Force to date indicate that there is no inherent reason why an environmental auditing program cannot be successful within its decentralized management philosophy.

Those elements that are critical to the success of such a program are supplied by people and not necessarily by staff position. AES believes the capability of its people is second to none and wholly supports, even encourages, their stewardship of specific projects and tasks with whatever resources are necessary to successfully complete the endeavor.

The benefit of AES's Environmental Task Force is already evident, if perhaps not totally realized. There is a greater overall appreciation and understanding of compliance issues. The individual knowledge of those persons who have served on audit teams has been greatly enhanced. Finally, the overall communications and networking value of this entire exercise cannot be underestimated.

References

1. The AES Corporation, % *Convertible Subordinated Debentures Due 2002,* Donaldson, Lufkin & Jenrette, J. P. Morgan Securities, Inc.; (Unterberg Harris; Prospectus, March 1992).

4

Environmental and Social Accounting

Twin Approaches for Measuring Sustainability in Business

David Wheeler

General Manager,
Ethical Audit,
The Body Shop International

Introduction

The Body Shop first committed itself to an active program of integrated ethical auditing at the beginning of 1994.[1] This followed successful experience in implementing audit programs for environmental protection and health and safety at work, and supplier screening programs for animal protection. In the case of environmental audits, The Body Shop elected in 1991 to follow the European Union Eco-Management and Audit Regulation[2] as the most rigorous, comprehensive, and rational framework available.[3] The company published three independently verified environmental statements in 1992, 1993, and 1994.[4] In each case the verifiers confirmed that the statements satisfied the requirements of the European regulation; this requires validation of data as well as the management systems and programs for delivering environmental improvements.

The development of environmental auditing and management systems at The Body Shop has been described elsewhere.[5] Nevertheless, it is worthwhile placing the environmental programs in their organizational and philosophical context because this is quite pertinent to the later development of social and ethical auditing.

Throughout the 1990s there has been increasing emphasis on the need for transparency and especially public disclosure of environmental impacts by industry.[6] In addition, the relevance of environmental management and auditing systems to wider issues of socially responsible business behavior is now becoming understood. Thus in 1992, the United Nations Conference on Environment and Development (UNCED) was able to produce a document (Agenda 21) which made explicit the importance of environmental management, auditing, and public disclosure of environmental impacts to the broader goal of sustainable development.[7] In preparation for the UNCED conference, Schmidheiny and others[8] produced a quite radical prescription for altering business behavior to take into full account the need to balance environmental conservation and economic development. This analysis has stimulated work in the Business Council for Sustainable Development (BCSD) and elsewhere on issues such as eco-efficiency and ecological tax reform.[9] It has produced some resonance in progressive business circles and in more forward-thinking national and international agencies.[10–12]

It is only recently, however, that the business management implications of the wider agenda have become apparent. Thus, heads of corporations and business leaders still talk about sustainable development when they really mean environmental protection and the conservation of natural resources.[13,14] Companies whose activities could hardly be less sustainable act as leading players in international forums promoting environmental management initiatives which are claimed to be compatible with unrestricted free trade and patently unsustainable consumption.[15] Even terms such as eco-efficiency, life cycle assessment, and eco-management are promoted as technical devices devoid of wider issues of global social responsibility and intergenerational equity.[16]

There is a need, therefore, to pioneer management systems which are based on a wider perspective of how environmental protection and conservation relate to business obligations. If we are to avoid the kind of crisis foreseen by some[17] and move toward a more genuinely "eco-centric" approach to business, new techniques are needed. These techniques may borrow from existing management theory and practice, including techniques used in environmental, health and safety, or quality management and auditing. But businesses will need a broader and more holistic set of values together with systems for implementing those values if they are to make a genuine commitment to sustainable development. A paradigm shift is needed, complete with methodological underpinning.[18,19] Integrated ethical auditing which takes into account social, animal welfare, and environmental protection issues is one technique which may help.[20,21]

This chapter describes in detail The Body Shop's approach to ethical accountability. The Body Shop's business activities are described, together with the company's ethical policies and how these are implemented and audited. Sections dealing with environmental protection, social issues, and animal protection are presented, and emphasis is given to the importance of auditing and disclosure of performance as the keys to establishing stakeholder support.

In January 1996, The Body Shop published an integrated statement of its ethical performance, called *The Values Report 1995*. Each component of the report has an element of independent verification in line with established best practice where this exists. This chapter will form part of the *Values Report* as a published guide to The Body Shop's approach.

The Body Shop as a Manufacturer, Distributor, Retailer, and Discloser of Ethical Performance

The principal subject of our *1995 Values Report* is our wholly owned business in the United Kingdom. Most of the information relates directly to our Watersmead (U.K.) site, home to The Body Shop's manufacturing, production, and warehousing, its international head office, laboratories (including research and development), a family center, a visitors' center, and a wastewater treatment plant. There are four components to our *Values Report,* namely this guide to our auditing and disclosure practices, plus three separate statements dealing with our ethical performance, titled Environment, Animal Protection, and Social Issues.

The Environmental Statement covers issues at our principal operating sites in Watersmead (Littlehampton, U.K.), Wick (Littlehampton, U.K.), and Wake Forest (North Carolina). In all three cases our environmental management systems and data have been subject to external verification in line with the provisions of the EU Eco-Management and Audit

Scheme (EMAS). Details of environmental performance at Soapworks, our Easterhouse soap factory (Glasgow, Scotland), and in our international retail markets are included for completeness but these have not yet been subject to external verification processes.

The Animal Protection Statement covers systems and procedures at Watersmead where the majority of product and ingredient purchasing decisions were made during the audit year. Procedures for checking the animal protection credentials of our suppliers are partially decentralized to purchasing departments on subsidiary sites and in two franchise operations (Canada and Australia). However, these activities are overseen by systems which operate at the Head Office level. Thus, independent verification of activities in 1994–1995 concentrates on the Watersmead audit and management systems.

Finally, the Social Statement focuses on our impacts on stakeholders directly affected by our U.K. operations. It has been possible to include some analysis of U.S. franchisees' views and preparatory work for more detailed work in the United States has been undertaken. However, most of the information in the Social Statement is focused on the performance of The Body Shop International.

In arranging our *Values Report* and ethical statements in this way we believe we have systematically covered most relevant issues as far as environmental protection, animal protection, and social issues are concerned. Naturally, the scope of each report needs to be fully understood. For example, as noted above, this year we were unable to include quantitative information from staff employed directly by our subsidiary company in the United States in our Social Statement, although we have initiated the process which will allow us to do that in future years. Similarly, because it is a relatively small facility, we have not yet included our soap factory's environmental management program in independent verification, although it is covered by our internal ethical audit systems. Below is a framework to help the reader understand the scope of the *Values Report* and the extent to which information in each statement has been subject to internal audit and external verification.

From Table 4-1 it is clear that most significant areas of the company are being addressed for all three areas of ethical concern in the United Kingdom. Further progress needs to be made in the United States—especially with respect to environmental information gathering and auditing in the company stores. In addition, appropriate social audit procedures need to be developed and put in place for our U.S. operations. 1996 will see tangible progress in both areas. Programs for international franchisees necessarily take more time to develop, but we have been very pleased with the environmental auditing and information gathering activities of franchisees since 1992. This will be further con-

Table 4-1. Scope of Audit and Verification Activities for Main Sites and Activities of The Body Shop International

	Watersmead site (U.K.)	Wick site (U.K.)	Wake Forest site (U.S.)	Easterhouse site (Scotland)	U.K. company stores	U.S. company stores	International franchises	U.K. franchises	U.S. franchises
Staff employed as of February 28, 1995	863	367	240	150	669	989	5500*	3200*	1595*
Environmental management & audit procedures	XXX	XXX	XXX	XX	XX	—	X(X)	XX	—
Animal protection & audit procedures	XXX	XX	XX	XX	N/A	N/A	XX	N/A	N/A
Social audit procedures	XXX	XXX	X	XXX	XXX	—	XXX	XXX	XXX

Key:

XXX = Information presented, internally audited, and externally verified.

XX = Information presented, internally audited.

X(X) = Information presented, internally analyzed.

X = Information presented.

— = No information presented.

N/A = Not applicable or relevant.

* = Estimated number of staff based on average number of staff employed in company stores.

Note: Table includes all areas of the company employing more than 100 employees.

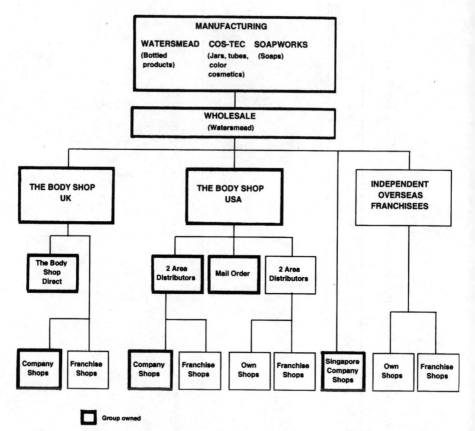

Figure 4-1. Operational structure of The Body Shop International as of February 28, 1995.

solidated during 1995–1996 and appropriate social audit programs introduced as soon as practicable.

In 1994, The Body Shop embarked on a fundamental reorganization of its activities and management systems with assistance from the Adizes Institute—a U.S.-based consultancy. The restructuring was largely complete by early 1995. The operational structure pictured in Fig. 4-1 was the one in place as of February 28, 1995.

The Watersmead Site (Littlehampton, U.K.) is the main headquarters, and manufacturing and distribution center for The Body Shop's international business. Our manufacturing and production operations at Watersmead involve hot and cold mixing of shampoos and conditioners, lotions and moisturizers, gels and cleansers, and oils and bubble bath products. We also blow and injection-mold our own bottles, jars, and caps on site. These are then filled, packed, and stored in the adjacent warehouse, which is the principal storage and dispatch point for goods worldwide.

The plant at the Watersmead site includes nine stainless steel mixing vessels (four cold mix and five hot mix), 13 filling lines, and, in the plastics department, eight blow-molding, six injection molding, and three injection blow-molding machines.

In addition to the bottled products listed above, the Watersmead manufacturing facilities provide bulk product filling into tubs, jars, and tubes by third-party contractors. Bulk concentrate and finished product is also shipped overseas for filling in the United States, Canada, and Australia. Products filled or manufactured by third-party contractors (including sundry and accessory items) are delivered to the Watersmead warehouse.

An increase in in-house bulk manufacturing on the Watersmead site was noted in 1994–1995 compared with 1993–1994. This can be attributed to increased production for overseas growth and increased in-house versus third-party manufacturing. The total in-house bulk manufacturing for 1994–1995 was 8066 tonnes compared with 7361 tonnes in 1993–1994 (+9.6 percent). Meanwhile 44.3 million bottles were filled and total issues from the warehouse were 15.7 million unit packs. These figures compare with 39.6 million bottles filled and 11.1 million unit packs distributed in 1993–1994 (+11.9 and +41.4 percent, respectively). In 1994–1995 we produced over 113.4 million plastic units compared with approximately 104 million units in 1993–1994 (approximately +9 percent).

Kids Unlimited operates a Family Center on the Watersmead site. It is for use by children of The Body Shop staff both at the Watersmead and Wick sites and has been built to provide child care for up to 60 children, ages ranging from three months to five years. The total number of child days in 1994–1995 was 12,429 compared with 11,123 in 1993–1994 (+11.7 percent). The afterschool scheme caters to children between the ages of 5 and 11 and has a daily attendance of 14 children who are collected from the local schools. On May 6, 1994 a family-care scheme was started to provide financial support for non-Littlehampton staff and Head Office staff who do not use the nursery.

The Body Shop first opened its doors to the public in November 1992, with a new Visitors Center at Watersmead. Tours are now run for approximately 300 people per day. The Trading Post houses a factory shop offering trade bargains, and also has a fresh fruit juice and snack bar. During the 1994–1995 audit period, 77,087 people made the trip to Watersmead for The Body Shop tour compared to 64,270 for 1993–1994, an increase of 19.9 percent. We have recently introduced environmental tours of the site specifically designed for school children doing projects. A wind turbine provides energy for the Visitors Center and is capable of supplying electrical power for the tour vehicles.

A wastewater treatment plant using an ultrafiltration system installed in August 1991 is also situated on the Watersmead site in a traditional Sussex barn. In addition, an experimental ecological wastewater treatment system has been in place since 1992. This is housed in a greenhouse adjacent to the Sussex barn. A full-scale ecological treatment system was commissioned to be completed in the winter of 1995, in order to ensure full biological treatment of all the wastewater from the ultrafiltration plant at Watersmead. The total quantity of wastewater treated in 1994–1995 was 7920 m^3; this compares with 5366 m^3 in 1993–1994 (+47.6 percent).

The Wick site was previously the Head Office of The Body Shop International but it now houses The Body Shop Colour Division manufacturing facilities along with several of the corporate departments. The Colour Division was formerly a subsidiary company named Cos-tec which was formed in July 1988. In 1991 the company was bought by The Body Shop and was relocated from its original site in Burgess Hill to Wick. The Burgess Hill site is still in use on a small scale. The Colour Division manufactures about 60 percent of the Colorings color cosmetics range, together with a variety of toiletries including talc, foot care products, hair care products, and deodorants. The filling equipment is known for its versatility allowing the filling of products such as aromatherapy oils and potpourri. The plant includes nine mixing vessels, seven lipstick vessels, two pulverizers, four ribbon blenders (for powders), and three vats for alcohol-based products. There are 24 production lines producing approximately 750,000 finished units per week.

Soapworks, which produces the majority of our soaps sold in shops around the world, is based in Easterhouse (Glasgow, Scotland). Plant at Soapworks includes three soap lines, seven wrapping machines, one "Play Soap" mixer, and a sachet machine. In 1994–1995, Soapworks manufactured 35 million soap bars, an increase of around 7 million compared to the previous year.

Our design studio is based in Central London together with our retail operations training school. On February 28, 1995, we had 143 company shops located worldwide—43 in the United Kingdom, 91 in the United States, and 9 in Singapore. There were 21 "partnership" shops run by staff in the United Kingdom and 179 shops were franchised. Around the world the total number of shops on February 28, 1995 was 1213. The number of markets we trade in now totals 45 and the number of languages we deal in is 23.

It is now two years since The Body Shop in the United States gained a new home in Wake Forest, North Carolina. The total number of staff employed at the Wake Forest site at the end of February 1995 was 240 (compared to 130 staff the previous year) including office, warehouse, and production personnel. No products are actually manufactured in the United States, but bulk concentrates are received at Wake Forest where they are filled into bottles which are purchased locally. There are three automated filling lines and one manual line at the plant. The total bottles filled in the United States from March 1, 1994 to February 28, 1995 amounted to 7,641,184. As with the Wick site, Wake Forest was included in the audit process, and systems and data externally verified for the first time this year.

There are also two fully-owned area distributorships in the United States—Cedar Knolls, New Jersey and Hayward, California. The total number of staff for Cedar Knolls and Hayward on February 28, 1995 was approximately 86 and 47, respectively. They have not been included in formal audit processes to date as they are relatively small facilities. Cedar Knolls was due for closure during 1996.

Our Approach to Ethical Auditing and Disclosure: Policies and Organizational Structure

The Body Shop's Mission Statement (Fig. 4-2) was adopted in mid-1994. It dedicates the company to the pursuit of social and environmental change. It is a holistic document, embracing human and civil rights, ecological sustainability, and animal welfare. The Mission Statement also makes clear that the company should work tirelessly "to narrow the gap between principle and practice." Underpinning the Mission Statement is a trading charter (Fig. 4-3) which addresses the three principal ethical concerns of The Body Shop while committing the company to "appropriate monitoring, auditing, and disclosure mechanisms to ensure our accountability and demonstrate compliance" with our trading principles.

Together with The Body Shop's Memorandum of Association, which also affirms the values-led nature of the business, the Mission Statement and Trading Charter provide the central thrust for the company's ethical policies and its desire to demonstrate accountability on ethical issues. At the present time, The Body Shop maintains a number of formal policies, guidelines, and procedures manuals which underpin the

---

Mission Statement—Our Reason for Being

- To dedicate our business to the pursuit of social and environmental change.

- To creatively balance the financial and human needs of our stakeholders: employees, customers, franchisees, suppliers, and shareholders.

- To courageously ensure that our business is ecologically sustainable: meeting the needs of the present without compromising the future.

- To meaningfully contribute to local, national, and international communities in which we trade, by adopting a code of conduct which ensures care, honesty, fairness, and respect.

- To passionately campaign for the protection of the environment, human and civil rights, and against animal testing within the cosmetics and toiletries industry.

- To tirelessly work to narrow the gap between principle and practice, while making fun, passion, and care part of our daily lives.

Figure 4-2. The Body Shop mission statement.

ideals expressed in the Mission Statement. Some are still under development and await formal release. However, the current picture is depicted in Fig. 4-4.

In early 1994, in order to promote compliance with its ethical policies, The Body Shop set up an integrated Ethical Audit Department which aims to provide support for the company's ethical stance in the following areas:

- Policy development and maintenance
- Auditing and reporting
- Advice and training

The Ethical Audit Department has six areas of professional expertise:

- Animal protection
- Environmental protection
- Social issues
- Health and safety at work

Our Trading Charter

The Way We Trade Creates Profits with Principles

We aim to achieve commercial success by meeting our customers' needs through the provision of high quality, good value products with exceptional service and relevant information which enables customers to make informed and responsible choices.

Our trading relationships of every kind—with customers, franchisees, and suppliers—will be commercially viable, mutually beneficial, and based on trust and respect.

Our Trading Principles Reflect our Core Values

We aim to ensure that human and civil rights, as set out in the Universal Declaration of Human Rights, are respected throughout our business activities. We will establish a framework based on this declaration to include criteria for workers' rights, embracing a safe, healthy working environment, fair wages, no discrimination on the basis of race, creed, gender, or sexual orientation, or physical coercion of any kind.

We will support long-term, sustainable relationships with communities in need. We will pay special attention to those minority groups, women, and disadvantaged peoples who are socially and economically marginalized.

We will use environmentally sustainable resources wherever technically and economically viable. Our purchasing will be based on a system of screening and investigation of the ecological credentials of our finished products, ingredients, packaging, and suppliers.

We will promote animal protection throughout our business activities. We are against animal testing in the cosmetics and toiletries industry. We will not test ingredients or products on animals, nor will we commission others to do so on our behalf. We will use our purchasing power to stop suppliers' animal testing.

We will institute appropriate monitoring, auditing, and disclosure mechanisms to ensure our accountability and demonstrate our compliance with these principles.

Figure 4-3. The Body Shop trading charter.

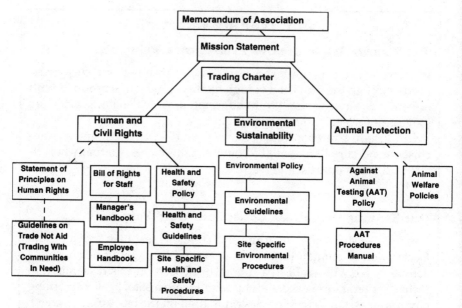

Figure 4-4. Guide to The Body Shop International key policies on ethical issues (dotted lines indicate where policies are still under development).

- Information management
- Training

The department is organized along stakeholder lines, with each of the main professional groupings taking special responsibility for the needs of particular stakeholders (see Fig. 4-5). This integrated matrix-style approach to ethical auditing has proven effective in streamlining communications and avoiding Head Office departments and stakeholders being confused by differing values-related audit demands.

The position of the Ethical Audit Department with respect to the rest of the company is shown below (Fig. 4-6). It may be noted that, as part of the Values and Vision Center, the department reports directly to founder and chief executive officer, Anita Roddick, and is represented on the Management Committee in the CEO's absence.

The Audit Department is not usually responsible for day-to-day operational management of ethical issues. Where it is necessary to develop a system of control centrally, the Ethical Audit Department may take provisional responsibility for it. But this is always with the intention of floating the system off into the relevant business entity at the earliest

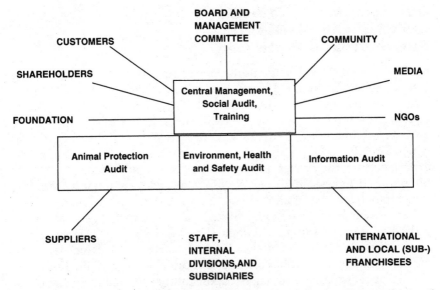

Figure 4-5. Structure of Ethical Audit Department showing how professional groups take responsibility for ethical issues of relevance to principal stakeholders. Each grouping also works through relevant departments (e.g., through purchasing groups for supplier issues and through investor relations for shareholders).

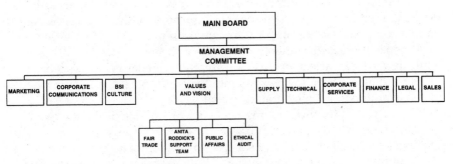

Figure 4-6. Structure of The Body Shop International principal divisions showing detail of Values & Vision Center.

appropriate opportunity. Thus, for example, the Supply Division maintains its own Health, Safety, and Environmental management group, the Purchasing group has its own systems for moving suppliers forward on ethical issues, and the Technical Division maintains its own systems and records on the ethical profile of product ingredients.

Our Approach to Ethical Auditing and Disclosure: Environmental Protection

Environmental reviews and audits have come a long way since the early 1980s. Pioneered in the United States and largely driven by heavy industries that had to minimize civil liabilities, early audits tended to equate environmental risks with health and safety ones. This was a natural development for industries such as energy and petrochemicals which already had long track records of health and safety auditing. The experiences of the Soviet authorities in Chernobyl, Union Carbide in Bhopal, and Icmesa in Seveso certainly served to emphasize the relationship between occupational and environmental risks.

But growing environmental awareness among the general public prompted a growth in environmental auditing in companies who had lower levels of exposure to risks of industrial accidents. Thus, new techniques reflect wider public concerns and often go beyond the limitation of risks or compliance with legal responsibilities for health and safety or pollution control.

Audits now examine how to save money on waste treatment and disposal and how to reduce energy consumption; both are especially relevant for organizations that want to justify spending money on improved environmental performance. Audits can also cover procurement policy (including suppliers), global environmental responsibilities (e.g., the reduction of CO_2 emissions), education and awareness raising, and a company's whole approach to environmental strategy.

Because these issues often strike at the very heart of an organization's culture and can affect profits—positively *or* negatively—it's not surprising that methodologies available for environmental auditing have been examined in some detail in recent years.

Happily, despite the growing diversity of organizations involved in auditing, standards and approaches have been converging in recent years. Auditing is now seen as a tool to be employed within a formal environmental management system (EMS). An EMS establishes effective ways of detecting and responding to environmental problems.

In late 1991, The Body Shop International adopted the (then) draft European Community [now European Union (EU)] Eco-Audit Regulation as the principal framework for the company's environmental maagement, auditing, and public reporting. The first Environmental Statement was published in May 1992, focusing particularly on the environmental performance of the main headquarters and manufacturing site at Watersmead, Littlehampton, United Kingdom.

During 1992–1993, the draft Eco-Audit Regulation underwent further negotiation and development, eventually emerging as the European Union Eco-Management and Audit Scheme (EMAS). The scheme retained its voluntary nature, a point of some concern to The Body Shop. But several essential components, most notably continuous improvement of performance, a commitment to best practice, and independently verified public disclosure, ensured that the measure retained credibility. Accordingly, The Body Shop published two further environmental statements in line with EMAS in May 1993 and June 1994. Again, the statements concentrated on environmental performance on the main site, but information was included from other parts of the business for the sake of completeness. In 1995, The Body Shop extended independent verification and reporting to its other main sites at Wick (Littlehampton, United Kingdom) and Wake Forest (North Carolina), thus expanding the scope of its verified public reporting. In addition, data provided by international franchisees were collected and analyzed for the first time and included in the public statement.

Since the formalization of internal environmental management systems and audit programs in 1991–1992, The Body Shop has relied almost entirely on its in-house professionals rather than external consultants. External consultants have only been employed for specialist inputs to environmental programs, e.g., energy management, waste minimization, and verification. In our case this has proven cost-effective and efficient. The period 1991 to 1995 has also seen a steady consolidation of the split between day-to-day site-based environmental management (which is now conducted on behalf of the company on a site-by-site basis by the Global Supply Division) and the audit function. This split, which may be likened to the difference between quality control and quality assurance, avoids potential conflicts of interest and ensures maximum integration of environmental programs within the business.

The Body Shop's experience of EMAS has been very positive. Despite the parallel emergence of British (BS 7750) and International (ISO 14001) standards on environmental management systems, we believe that EMAS represents by far the most exacting framework for ensuring the best environmental practice in industry. Table 4-2 gives a comparison of the requirements of EMAS alongside other standards, including some voluntary codes of practice advocated by a number of industry bodies. The table is reproduced by kind permission of Environmental Resources Management (ERM). It illustrates why EMAS remains the most demanding environmental standard today, particularly with respect to public disclosures.

Figure 4-7 illustrates the key requirements of EMAS. It should only be

Table 4-2. Management System Components[*]

System component	ISO 14001	EMAS	BS 7750	ICC	PERI	CEFIC	Keidanren
Company policy	XXX	XXX	XXX	XX	X	X	X
Senior management commitment	XXX	XXX	XXX	XX	X		X
Initial review of impacts	XXX	XXX	X	X	X	X	X
Register of regulations	1	XXX	XXX	2	X		
Register of significant impacts	1	XXX	XXX		X		
Allocated responsibilities	XXX	XXX	XXX	XX	X		X
Objectives and targets	XXX	XXX	XXX	XX	X	X	X
Management program	XXX	XXX	XXX	XX	X	XX	X
Manual	XXX	XXX	XXX				
Operational control	XXX	XXX	XXX	XX			X
Records	XXX	XXX	XXX				
Training	XXX	XXX	XXX	XX	X	X	X
Audits (internal)	XXX	XXX	XXX	XX	X	X	X
Public statement and reporting	X	XXX		XX	XX	X	X
System verification	XXX	XXX	XXX			X	
Statement and report verification		XXX				X	
Commitment to continuous improvement of system	XXX	XXX	XXX	XX		XX	

Key:
XXX = A requirement of the standard.
XX = Principle.
X = Guideline/good practice.
1 = Review regulations/impacts.
2 = Assess compliance.
Reproduced by permission of Environmental Resources Management.

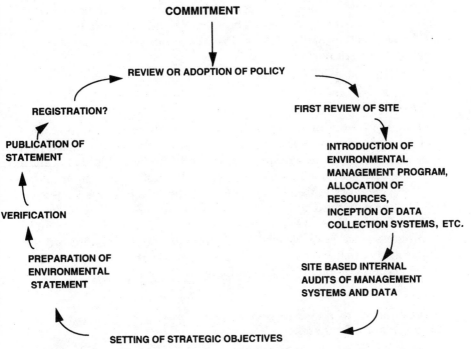

Figure 4-7. Framework for environmental management, auditing, and disclosure at The Body Shop (based on main requirements of the European Union Eco-Management and Audit Scheme).

a matter of applying logic for any industrial concern to follow each step in the loop to the satisfaction of external verifiers in order to meet the standard. However, in the experience of The Body Shop, logic is only half the story. The other half is hard work and commitment throughout the organization. Like any management system, it only works effectively if the organization itself is dedicated to adoption at every level.

Listed below are the key requirements of EMAS with commentary on how and when The Body Shop put in place the components necessary to meet the standard.

Review or Adoption of Policy

The Body Shop adopted environmental goals for the first time in 1990. These were based loosely on the Valdez (now CERES) principles of corporate environmental accountability and were designed in a series of meetings involving senior managers of the Company. In early 1992, these goals were converted into a formal policy, endorsed by the main board of The Body Shop. They committed the company to best practice and continuous improvement. The main policy was next updated in 1994 and issued formally, along with a new comprehensive set of guidelines and procedures in early 1995.

First Review of Site

Although an "environmental audit" of The Body Shop was conducted in 1989 when the company was still based at the Wick site, in retrospect this would now be described as a "review" because it did not audit against policies or management systems. The first formal environmental assessment of the Watersmead site was conducted in late 1991–early 1992 in line with the draft provisions of the EC Eco-Audit Regulation. So it is a debatable point whether this constituted a first review or an audit of progress since 1989. Nevertheless, the external verifiers did confirm in April 1992 that the 1991–1992 assessment did at least meet the first review criteria of the draft EC regulation.

Introduction of Environmental Management Program

In preparation for the 1991–1992 assessment of environmental performance on the Watersmead site, individual departmental objectives were set throughout the organization. These took the form of terms of reference for Departmental Environmental Advisers (DEAs)—a volunteer network of ordinary staff members. Departmental and DEA objectives were agreed with each departmental manager in a series of meetings which served both to formalize the DEA role and to set out the departmental agenda for the following year. In addition to the departmental objectives, a number of strategic targets were put together with time scales for implementation. These were published in our first environmental statement: *The Green Book.* The document was published alongside the annual report and accounts—that year described as *The Black Book.*

Site-Based Internal Environmental Audits

In 1992, The Body Shop's corporate Environment, Health and Safety Audit Department designed management systems audit checklists for application throughout the organization. These checklists evolved through an examination of best practice elsewhere, through consideration of the requirements of the European regulation, and through an analysis of issues and priorities relevant to the company's international business. In the latter case, much useful input was obtained from a group of international franchisees convened specifically for the purpose of helping shape the future environmental audit procedures of The Body Shop. The audit checklists which emerged were applied for the first time in the autumn of 1992 at the Soapworks site in Glasgow and were subsequently used on all main operating sites of The Body Shop in the United Kingdom and United States. The internal management systems audits based on these checklists took the form of face-to-face confidential interviews throughout the organization. Checklists were analyzed quantitatively and qualitatively, and internal management reports were filed with each part of the organization audited. Since 1993, audit checklists have been updated and developed, but the basic format has remained consistent. Thus, despite the reorganizations which occurred between 1992 and 1994, most parts of the business have received two internal audits which permit both quantitative and qualitative assessment of progress. Reports based on these audits are submitted to divisional and departmental managers who then have the responsibility to implement recommendations. These reports are not published.

Setting Strategic Objectives

Since 1992, departmental objectives have continued to be set on an annual basis for most Watersmead-based activities. In 1993, these were produced in a stand-alone format (*Green Book 2: The Detail*) which was published internally as an awareness-raising document. In each case departmental objectives were formally signed off by senior managers and their departmental environmental adviser. In 1994, objective setting at Watersmead was devolved to operating divisions. The corporate departments continued to produce a formal document for internal awareness raising, but other divisions relied more on internal memoranda. From 1992 to 1995 the recommendations of the independent verifiers were taken up automatically into the next round of departmental and strategic objective setting, but in some cases this led to a delay because of inopportune timing. So, in 1995, final objective setting was left until after the receipt of the detailed reports of the external verifier, to ensure that any recommendations made as part of the internal audit or external verification process could be immediately and automatically embraced as part of the 1995–1996 environmental program for each division. In addition to internal objective setting, strategic company-wide targets have continued to be set on an annual basis and published in the Environmental Statement.

Preparation of Environmental Statement

1996 sees publication of The Body Shop's fourth environmental statement. The verified section of the Statement now includes data from all three principal manufacturing sites of The Body Shop. In the nonverified section (included for completeness), data are included from international markets for the first time; there are also small sections on Soapworks, the Glasgow-based soap factory, and the U.K. retail stores. Every year, the draft statement is available to the external verifier prior to commencement of their verification of the environmental policies, programs, management systems, and internal audits. Following advice from the verifier, minor alterations are usually made to the Statement to improve its public understanding.

Verification

Each year since 1992, The Body Shop has commissioned external verifiers to attest to the fact that the company's site-based environmental management and reports meet the criteria of the European Union Eco-Audit (now Eco-Management and Audit Scheme—EMAS). As noted above, in the first three years this verification covered the Watersmead site only. In 1995, this was extended to include the Wick and Wake Forest sites. In each case the verifiers (Environmental Resources Management) spend several person-days on-site testing the efficacy of procedures and the accuracy of the environmental statement. Following the physical verification, a formal verification statement is submitted for inclusion with the public environmental statement. A longer report, which includes confirmation of the status of management systems with respect to EMAS, together with more detailed recommendations for improvement, is also submitted by the verifiers.

Publication of Environmental Statement

Following verification, the Environmental Statement is published and distributed to interested parties. In 1992, a summary broadsheet was also distributed alongside the full Environmental Statement. This proved especially popular with retail stores and customers. Similar summaries and popularizing mechanisms have been put in place for the 1995 Statement.

Registration

Although the European Regulation has been technically available for companies to comply with and register since April 1995, member states of the EU are only now putting in place formal systems for accrediting verifiers and registering participating industrial sites under EMAS. Thus it was not possible for the 1995 Statement to be formally registered with the national competent body in the United Kingdom and the European Commission. The Body Shop will consider whether there is any value in formal registration when arrangements are finalized.

Seven Dos and Don'ts of Environmental Auditing and Disclosure

Do get free or cheap advice wherever it is available: DIY manuals, local and central government departments local and national business clubs, simple booklets and guides to the issues.

Don't assume that you have to be an expert. A great deal of environmental management and auditing is simply common sense. Networking and asking questions of similar businesses is the best way to avoid mistakes.

Do set up informal networks of internal supporters and environmental representatives to aid communications and internal campaigns.

Don't put too many resources into auditing if they are better directed at obvious priority actions and improvements, especially in the early days.

Do involve departments, managers, and staff at every level, especially in policy formulation, internal goal setting, communications and training.

Don't forget that people want to be inspired by a vision; environmental activists are often happy to talk to busness groups and companies committed to improving their environmental performance and participate in external campaigns.

Do set up an internal independent audit system or department and have them report to a main board director.

Don't forget to motivate train, and then remotivate your managers, staff, and volunteer networks.

Do use consultants for specific tasks, e.g., first review, audit verification, technical tasks (e.g., energy auditing). Always network with other businesses to find recommended consultants or local academic institutions for support.

Don't use consultants for general activities that are the proper responsibility of staff anf managers, e. g., objective setting, resource allocation, coordination. If management cannot take these basics on board you are wasting your time.

Do consider adopting a formal standard for environmental auditing and management, but only when you're ready to commit.

Do report: formally and informally, publicly and internally. Stakeholder understanding is crucial to progress, as are targets and objectives for the future.

Don't kill enthusiasm by introducing too many systems too early. Make it fun and don't be afraid to be controversial.

Don't be afraid of including both good and bad aspects of environmental performance; better that you draw attention to your faults than have your critics do it.

Our Approach to Ethical Auditing and Disclosure: Social Auditing

The development of social auditing follows a long tradition of exploring the social and ethical dimension of business behavior. Early references to social auditing can be traced back to the late 1960s. The first attempts were largely concerned with expanding conventional financial reporting to include information on "corporate social expenditure." In the United States, social reporting practices were rapidly adopted by large corporations as a response to intensifying criticism by shareholders and the general public. By the late 1970s the majority of Fortune 500 companies included a page or two in their annual reports to inform readers of their "social expenditure" or to disclose statistical information on, e.g., equal opportunities or related issues.

A second type of social auditing developed in the 1980s. This new wave was essentially driven by the consumer movement and by the evolving ethical investment community who were keen to build up and publish profiles of corporate social performance. Although the majority of these profiles were not referred to as social audits or reports, that was very much what they were trying to do—to inform customers or other interest groups about how the assessed companies rated against certain sets of social and ethical criteria. The fact that a number of these investigations became instant best-sellers in the late 1980s and early 1990s— for instance, *Shopping For A Better World* and the *100 Best Companies to Work for in America*—demonstrated that there was great interest in such information.

The late 1990s have seen the beginnings of a third wave in social auditing driven by companies who have adopted an active moral stance. This time social auditing is not a process imposed from the outside; it is driven by the organization itself.

In 1991, Traidcraft Plc, a U.K.-based fair trading organization, joined forces with the not-for-profit research organization, the New Economics Foundation (NEF), to develop a more systematic approach to social auditing. The resulting audit method was born out of a mix of community-based participative research techniques and organizational development literature. At its core was a "stakeholder approach." Traidcraft has now released several social accounts. Its initiative is perceived as a pioneering effort to advance more rigorous approaches to social accounting.

Another manifestation of the internally driven social audit was the Ethical Accounting Statement developed in Denmark by the Copenhagen Business School. This approach was first tried in 1990 by a regional Danish bank, Sbn Bank, which has since produced annual statements of its social performance. A number of Danish organizations, ranging from small companies to hospitals and schools, have now adopted this method.

Like the method developed by NEF and Traidcraft, the Ethical Accounting Statement involves a consultation process with the organization's key stakeholders. Beyond this, the ethical accounting approach encourages stakeholders to propose practical changes to the organization's operational practices; e.g., customer services or product development. Unlike the approach of NEF and Traidcraft, however, the published statement is not externally verified by independent auditors.

One further example characterizing the 1990s attempts to develop internally driven social reporting mechanisms is the one adopted by the American ice cream manufacturer, Ben & Jerry's. Like the above approaches, it tries to describe the social impact of the organization's activities on key stakeholder groups. Similar to the verified social audit and ethical accounting statements, it involves stakeholder consultation and external validation, albeit in a less systematized way. The main feature of Ben and Jerry's approach has been to invite a high profile advocate of corporate social responsibility to spend time each year exploring any aspect of the company's activities that he or she might deem important. The idea is that the external commentator has free access to the company's internal records and freedom to consult the company's stakeholders on an ad hoc basis. Based on this process the person then writes a personal evaluation or social statement.

As the New Economics Foundation has pointed out, a comparison of different approaches to social reporting is not straightforward, since the circumstances of different accounting processes are often critical in determining the design. The Copenhagen Business School Ethical Accounting Statement approach, for example, does not include an external verification process. However, verification may not be seen as crucial in the Danish context as it would be in socially less cohesive soci-

Table 4-3. Comparison of Different Approaches to Social and Ethical Accountability

Principle	Social audit	Ethical accounting	Social evaluation
Multiple stakeholder perspective	Yes	Yes	Yes
Comprehensive and systematic	Yes	Yes	No
Regular	Yes	Yes	Yes
Internal bookkeeping of social performance indicators	Yes	No	Yes
Systematic bookkeeping of stakeholder accounts	Yes	Yes	Yes
Independent external verification	Yes	No	Yes
Public disclosure	Yes	Yes	Yes

eties such as the United Kingdom. So there are some dangers in simplified comparisons between the methods. A deeper analysis of the different cultural contexts in which each of them has been developed and applied is required. With this qualification in mind, Table 4-3 summarizes some key aspects of these internally driven approaches.

Although there are plans to set up an Institute for Social and Ethical Accountability, there are, as yet, no universally agreed standards or frameworks for social auditing, there is no accepted terminology, and there are very few sources of expertise to call on.

The framework presented in Fig. 4-8 is therefore tentative. It is the framework which is currently used by The Body Shop. It endeavors to synthesize some aspects of environmental management and auditing systems with recent experience of leading practitioners in social auditing and disclosure.

Commitment

The Body Shop committed itself to a formal social audit process in 1992. This followed publication of Anita Roddick's autobiography in 1991 (*Body and Soul*) in which she wrote:

> I would love it if every shareholder of every company wrote a letter every time they received a company's annual report and accounts. I would like them to say something like, "OK, that's fine, very good. But where are the details of your environmental audit? Where are the details of your accounting to the community? Where is your social audit?

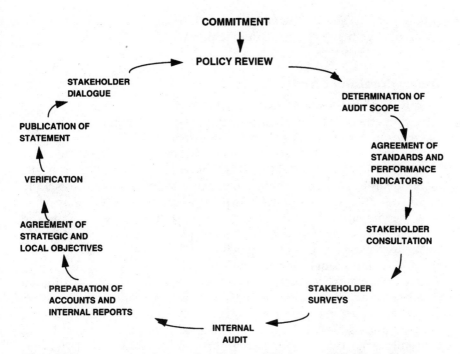

Figure 4-8. Framework for social auditing and disclosure at The Body Shop.

There is no doubt that had it not been for the commitment of the founders of the company and the development of an "accountability ethos" within The Body Shop in the early 1990s, the company would not now be in a position to publish a statement of its social performance. As with environmental management and auditing, clear leadership is probably the most important single factor in driving a process of social auditing and disclosure to a successful conclusion.

Policy Review

For any audit process to be meaningful, it is essential to understand which policy statements are relevant and which need to be developed in order that the organization's performance can be measured against a set of strategic goals. For The Body Shop's first social audit, the main policies against which performance might be judged were the company Mission Statement and Trading Charter. More specific policies and guidelines existed for health and safety at work, human resources (managers' and staff handbooks) and "trade, not aid" (trading with communities in need). These were also used as the basis for assessing the ability of management systems to deliver on policy commitments.

Determination of Audit Scope

Because parallel systems exist at The Body Shop for auditing and reporting on ethical performance with respect to animal protection and the environment, the subject area of the social audit was restricted to people: human stakeholders who may affect or be affected by The Body Shop. The number of individual stakeholder groups could theoretically be quite large and a decision has to be taken as to which groups should be included in the first and subsequent audit cycles. The Body Shop took the view that in the first cycle, the net should be cast as wide and as deep as possible. But inevitably some groups and subgroups could not be reached for practical purposes.

Another important factor in the scoping of a social audit is geography. Where a company has wholly or majority-owned operations in different countries, a decision has to be taken as to whether one or all countries are to be covered in each cycle. For The Body Shop this meant choosing not to include U.S. stakeholders in any great depth the first year but to concentrate mostly on U.K.-based groups. So, although some U.S. information was gathered with respect to franchisees and customers, it did not constitute quite the systematic effort that was devoted to the U.K. social audit process.

The last factor to be taken into account in scoping a social audit is the level to which indirect stakeholders may be embraced. For The Body Shop this required a decision about whether, for example, staff of franchisees or NGOs in franchised markets should be consulted. It was decided that such stakeholders were better consulted directly at such a time when franchisees were able to conduct their own audit processes.

Agreement of Standards and Performance Indicators

There are three types of performance measurement in The Body Shop's approach to social auditing. They are:

1. Performance against standards (benchmarks). These should reflect nationally and internationally available information on best practices for activities and policies that describe the organization's social performance. Measures may be both quantitative and qualitative. Standards are in agreement in memoranda of understanding with

relevant departments which then have the responsibility for collecting relevant information. Data are submitted by the departments and validated by the audit process. Policy and/or activity-specific audit checklists have been developed for use in internal audit interviews with relevant managers and staff.

2. Stakeholder perception of performance against core values (i.e., The Body Shop Mission Statement and Trading Charter). These core values are essentially defined by the organization itself. Each stakeholder group is surveyed to establish their perception of how closely the organization's social performance matches its stated values.

3. Stakeholder perception of performance against specific needs of stakeholders. These needs are particular to individual stakeholder groups. They are identified as salient through dialogue with stakeholders in focus groups and measured in market-research-style surveys.

Stakeholder Consultation

If social auditing is about giving a voice to stakeholders, one of the most important and sensitive processes is the engagement of stakeholder representatives in dialogue. It is especially important to identify salient issues for each group in face-to-face conversation before conducting surveys. The Body Shop has tended to use focus groups to allow stakeholder views and concerns to be expressed. For example, before conducting a wide-scale staff survey, 10 percent of the U.K.-based employees were involved in focus groups facilitated by the Ethical Audit Department. Focus groups excluded managers from The Body Shop who were responsible for those stakeholders participating in this discussion, in order to avoid inhibiting the free expression of views. In the case of one stakeholder group it was felt appropriate for the independent audit verifiers to facilitate a focus group. And in all cases the verifiers were invited to attend as observers. In this way the verifiers could be assured of fair play and open dialogue in the discussions.

Stakeholder Surveys

Following the focus groups, when specific issues had been identified as salient or of particular interest to stakeholders, questionnaires were designed to measure more wide-scale opinion. These questionnaires were intended to capture perceptions of the company's performance against both stakeholder-specific needs and core values articulated by the company.

Questionnaires were designed with professional assistance to avoid inadvertent introduction of bias. Space was also allowed on the questionnaires themselves for open-ended commentary on the company's performance.

Surveys were done using the largest manageable sample size; respondents completed the questionnaires anonymously and returned them to an independent survey organization for confidential analysis. Only summary information and lists of comments were submitted to The Body Shop for inclusion in the audit process. In some cases response rates were high (e.g., for staff who were given one hour off work to complete them) and in others quite low (e.g., for nongovernmental organizations who received the questionnaires by post). In most cases the resulting sample size was large enough to draw quite clear conclusions. In some cases the sample size was relatively small and care was needed in interpreting the results.

The Internal Audit

There were three main sources of information for The Body Shop audit process: (i) the results of the focus groups and surveys described above; (ii) the documentary information provided by departments which had agreed quantitative and qualitative standards; and (iii) the output from confidential interviews with staff and managers. This latter source of information was based on the kind of management systems structured interviews used in environmental auditing. Checklists were developed specifically for the purpose of the interviews, and results used to build up a dynamic picture of departmental and divisional handling and knowledge of social issues and company policies relevant to social performance. In future years it is intended to integrate management systems interviews from all The Body Shop's ethical audit procedures: health and safety at work, the social audit, environmental protection, animal protection, and information audit.

Preparation of Accounts (the Social Statement) and Internal Reports

To avoid the Social Statement becoming too densely packed with statistics or too discursive and "woolly," a balance has to be struck. Information from surveys has to be presented in a concise, user-friendly way and linking text has to avoid bias without becoming turgid. Statistics and benchmarks have to be balanced with quotes and views of stakeholders to bring the document alive. Finally, a sense of dialogue has to be created in which the company is seen to respond to stakeholder views, set out a direction, and commit to future progress on stakeholder relations.

The format chosen for The Body Shop's first Social Statement was based on a stakeholder model, with each group given its own section within the report. An introductory section gave a general explanation about the scope of the social audit and how the information had been compiled and what assumptions were used. The company founders also gave their overview in a foreword thereby setting a tone and direction to the document.

Each stakeholder section then followed a common format:

1. The basis for the company's approach and aims for each group (with reference to relevant policies, etc.)

2. The methodology used for each consultation process (i.e., what combination of focus groups, surveys, and discussions were used to capture stakeholder perceptions)

3. The results of stakeholder consultation with perception surveys described in as even-handed and neutral a way as possible so as to avoid premature interpretation, together with direct quotations from stakeholders selected in an independent fashion

4. Quantitative and qualitative standards of performance where these exist

5. A company response in the form of a quote from a Board member or senior manager setting out their reaction to the stakeholder views and noting where progress is already being made and/or where improvements are clearly required.

Setting out the accounts in this way promotes the dialogue process and allows stakeholders to take a view on the adequacy or otherwise of the company response. In order to make the follow-on dialogue process efficient, stakeholders receiving the Social Statement are encouraged to complete a response card and attend a discussion with representatives of the company.

The final components of The Body Shop's Social Statement were a verification statement from the New Economics Foundation, and a summary chapter on those stakeholders who could not be included in the cycle, but should be in future cycles. The external verification process should have an influence on the tone, format, and style of the Social Statement. The extent of this influence is guided by the verifier.

Agreement of Strategic and Local Objectives

As with environmental auditing and reporting, a very important part of the process is to set strategic objectives for the business which can help clarify the future priorities of the company and unite all stakeholders behind a common purpose. Setting strategic objectives and placing these in the public domain requires a significant amount of internal discussion, management commitment, and senior "sign off." It is not a process which can be rushed. Strategic objectives must also be underpinned by subsidiary or localized objectives which apply more specifically to individual stakeholder groups or parts of the company. The latter are in some ways simpler to negotiate because they involve fewer decision-makers at the corporate level. However, the endorsement and support of the company Board and central management committee are essential if more localized objectives are to be executed speedily and efficiently and kept in line with wider business goals.

Verification

Unlike verification of an environmental audit process and statement, where verification can safely be left toward the end of the cycle, social audit verification requires engagement throughout. The main reason for this is that when verifying a process aimed at an audit of human relations, it is not enough to examine documentation, conduct interviews with representative staff, and test the accuracy of data. The very process of collecting views from stakeholders has to be witnessed as fair and open.

So a social audit verifier has to engage and participate at every part of the cycle. This could result in the verifiers becoming compromised, i.e., too involved to be objective. So they engage, in turn, a formal advisory panel which meets to assess both the process and the output of the audit. The Body Shop's first social audit was verified by the New Economics Foundation; NEF appointed people with expertise in a wide range of social policy areas relevant to the company's audit. The panel of experts met on two occasions and contributed significantly to the framing of the Social Statement. NEF also produced a Verification Statement which summarized their view of the process and which included areas to be addressed in future audit cycles.

Publication of Statement

The Social Statement needs to be a true and fair picture of the social impacts of the organization, insofar as the defined scope of the audit allows. It needs to be comprehensive and systematic, but above all be understood. Because of the complex nature of a social audit and because of the variety of stakeholder needs for information, a multitier approach has been adopted for the publication of The Body Shop's first Social Statement. The full statement was published based on the approved, verified accounts. In addition, more detailed information was provided to staff on specific results relevant to them and their part of the company. A summary document of all the ethical statements (including the Social Statement) was produced for wider-scale distribution alongside even briefer material appropriate for customers and other large audiences. In each case readers were made aware of how they could obtain more detailed information.

Dialogue

Following publication of the results of the social audit, it is vital to obtain feedback from stakeholders and engage them in dialogue about how they react to the findings presented in the Social Statement. This process of dialogue helps shape future audit cycles, enables indicators and data presentation to be fine tuned for future cycles, and helps set priorities for future action by the company. Ideally, dialogue should be driven by those departments and divisions which have direct responsibility for stakeholder groups, but facilitated and attended by the audit team to ensure continuing openness of communication.

Seven Dos and Don'ts of Social Auditing and Disclosure

Do start with environmental auditing and disclosure if these are relevant to your organization. Environmental audits are simpler to organize and conduct than social audits.

Do consider joining the Institute for Social and Ethical Accountability—an important source of independent advice and experience.

Don't launch into a social audit without talking to someone who has done one. It is a long-term commitment, so plan ahead at least two audit cycles.

Don't forget the importance of training for social auditing: for managers and auditors. In its current form it is a new science and the principles and pitfalls need to be understood.

Do involve departments, managers, and staff at every level, especially in deciding the scope for the audit. Key departments are those that have most to do with stakeholder groups, e.g., human resources, communications/PR, investor relations, etc.

Do set up an internal audit system or department and have them report to a main board director.

Do exercise real care in selecting an independent verifier; they will have access to the very soul of the organization and their integrity is paramount. Always network to find verifiers with experience who are recommended by others.

Do allow plenty of time for drafting and finalizing the social statement. Audited departments will be very keen to be involved in putting results in context and proposing priorities for improvement.

Do report: formally and informally, publicly and internally. Stakeholder understanding is crucial to progress, as are targets and objectives for the future.

Don't forget to focus on the benefits and business case for social performance measurement and disclosure for all stakeholders. Good social auditing should make an organization more responsive and efficient.

Don't forget to publicize the role of the audit team and its purpose; people may feel more threatened by a social performance audit than by an environmental audit.

Don't forget that you may also need other sources of expert ad vice, e.g.,survey design and analy sis.

Don't allow one stakeholder voice to outweigh others. Take into account minority views but don't let them take over; a good external verifier will act as wise counsel on the right balance to be struck.

Don't be afraid of including both good and bad aspects of social performance; better that you draw attention to your faults than your critics do.

Our Approach to Ethical Auditing and Disclosure: Animal Protection

Organizations which conduct animal experiments are usually subject to some degree of external oversight. In most countries, licensing arrangements are handled by governmental bodies who seek to ensure the minimization of animal suffering during and after experimentation. In the United Kingdom licensing is handled by the Home Office. In addition scientists are often subject to professional or institutional codes of practice which also seek to ensure the humane treatment of test animals. Depending on the type of license, the nature of the experiments, and the purpose of the research, oversight might involve site visits by regulatory agencies.

For organizations which choose not to undertake or commission animal tests there is clearly no compulsion to obtain licenses or become subject to an external or internal professional audit of practices. However, in the cosmetics and toiletries industry, where the avoidance of animal cruelty embraces not just the company's own behavior but that of suppliers, some type of assessment is necessary.

Companies which market cosmetics are now well aware of the interest of consumers in avoiding cruelty to laboratory animals. Historically there have been four levels of company response to this consumer concern:

1. Some companies prefer to continue to test or commission tests either to provide arguments against liability claims should product safety ever be challenged or to satisfy third-party requests for such testing.

2. Some companies have found it possible to terminate the testing of finished products but reserve the right to continue to test, or ask suppliers to test, individual ingredients.

3. Some companies do not test or commission tests on animals either for finished goods or ingredients; however, they may tolerate using suppliers who do test raw materials used for their products, e.g., for regulatory or marketing reasons.

4. Some companies do not test or commission tests and use some kind of standard or purchasing rule against which to judge their suppliers and avoid encouraging the perpetuation of testing in the supply chain.

Over the years, animal welfare groups have recognized these different approaches by placing companies on approved or disapproved lists. These lists are almost universally based on information supplied by the companies themselves, and although the information is made public in order to help advise consumers, there can be no guarantee that a company is actually doing what it says it is doing on the animal testing question. This uncertainty has led to some confusion and public debate, particularly where animal welfare groups exercise different criteria for judging company behavior.

Recognizing this source of uncertainty, in 1994 The Body Shop started talking to animal welfare societies about how its own procedures and purchasing criteria could be subject to some type of external independent assessment. The idea was, if The Body Shop's systems (established in 1984) could be verified, then so could other companies' too. These discussions led to a decision in late 1994 to engage a quality systems assessor to check The Body Shop's procedures against an International Standards Organization (ISO) standard (ISO 9002). This standard is aimed particularly at assessing conformance to specified requirements by a company's suppliers; the logic is that companies can provide a better service or product quality to their customers if their suppliers are also performing adequately.

ISO 9002 is versatile enough to mean that any company with a non–animal testing stance which is actively seeking to check the performance of its suppliers could be assessed as another dimension of quality. The particular purchasing rule or standard is thus somewhat less important than the management systems and procedures which are in place to back it up. This finally overcomes the fairly futile debate over differing purchasing rules and should allow animal welfare groups to unite by classifying companies using a common, independently verified standard.

The scope of The Body Shop's assessment against ISO 9002 is "The monitoring of raw materials used in the products manufactured by The Body Shop International plc for compliance with company Against-Animal-Testing policies. The assessment and rating of the practices of raw material suppliers regarding animal testing."

The Body Shop was formally assessed against ISO 9002 in March 1995 and subsequently had compliance confirmed at a half year update in October 1995. This biannual independent check should continue in perpetuity.

Figure 4-9 sets out the key steps in The Body Shop's approach to compliance with ISO 9002 for its animal protection policies. All pertinent parts of the Standard have to be satisfied.

Review or Adoption of Policy

Since its inception, The Body Shop has a policy of not testing and not commissioning tests on animals for its products and raw materials. In the mid-1980s, this policy was extended to embrace suppliers so that there was no "blind eye" or indirect encouragement to animal test. This was achieved by introducing a system of regular declarations which suppliers had to complete for each raw material or product supplied to The Body Shop confirming compliance with purchasing requirements. The Body Shop's Against-Animal-Testing (AAT) policy was updated in mid-1994 and is due for updating again in 1996.

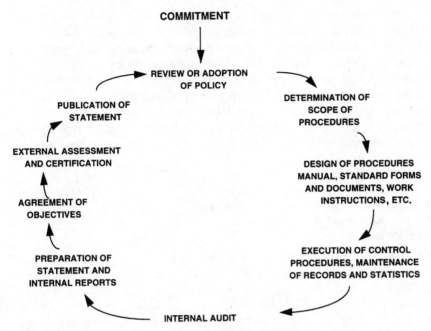

COMMITMENT

REVIEW OR ADOPTION
OF POLICY

PUBLICATION OF
STATEMENT

DETERMINATION OF
SCOPE OF
PROCEDURES

EXTERNAL ASSESSMENT
AND CERTIFICATION

DESIGN OF PROCEDURES
MANUAL, STANDARD FORMS
AND DOCUMENTS, WORK
INSTRUCTIONS, ETC.

AGREEMENT OF
OBJECTIVES

PREPARATION OF
STATEMENT AND
INTERNAL REPORTS

EXECUTION OF CONTROL
PROCEDURES, MAINTENANCE
OF RECORDS AND STATISTICS

INTERNAL AUDIT

Figure 4-9. Framework for Against-Animal-Testing auditing and disclosure at The Body Shop (based on requirements of ISO 9002).

Determination of Scope of Procedures

Most raw material purchasing decisions are taken at The Body Shop's main Head Office and manufacturing site at Watersmead, Littlehampton. On other manufacturing sites run by the Company (Wick, Soapworks, and Wake Forest) purchasing decisions are also subject to the AAT policy as they are at production sites run by franchisees in Canada and Australia. However, since the audit or quality assurance procedures operate from the Watersmead site, it is this site and its oversight of global purchasing functions which is the subject of ISO 9002 assessment.

Design of Procedures Manual, Standards Forms, and Other Documentation

The discipline of ensuring conformance to ISO 9002 meant that a number of "custom and practice" procedures had to be formally written and collated with existing quality control activities. Supplier non–animal testing declarations and in-depth supplier questionnaire forms were also consolidated and included with policies, organizational structures, and responsibilities and procedures in a single policy and procedures manual. All of this documentation is akin to a conventional quality manual; it simply deals with non–animal testing issues rather than the physical or technical quality of ingredients. The documents are laid out in a consistent format and style, their issue and updating is subject to formal control, and each is signed off by the senior managers responsible for the audit function.

Execution of Control Procedures, Maintenance of Records, and Statistics

Raw materials suppliers to The Body Shop have to complete regular declarations of conformance with the purchasing rule. Traditionally this QC-type function has operated from the Technical Division which is responsible for controlling all information relevant to formulations used for The Body Shop products. More recently, in-depth research into suppliers' activities on product safety testing in general has been conducted directly by the Ethical Audit Department. It is this research which has illuminated the situation on retesting of existing raw materials, and the commitment of suppliers to continued testing for regulatory and other reasons (i.e., just because suppliers do not test materials supplied to The Body Shop does not mean that they do not test other materials for other customers). Activities of suppliers with respect to alternatives to animal tests also formed part of the in-depth research.

The supplier nontesting declarations and the in-depth research generate statistics and records which must be carefully maintained and archived. In the case of the declarations, their careful maintenance was a crucial component of a successful libel case in 1993 when The Body Shop's activities in this area were challenged by a television program. In due course it is possible that day-to-day responsibility for pursuing in-depth research (which is similar to research conducted for environmental reasons) will transfer to purchasing groups as part of their overall approach to rating the performance of suppliers on ethical, technical, and business criteria.

Internal Audit

The Ethical Audit Department has protocols and checklists which it uses to audit compliance of manufacturing and production sites with The

Body Shop's AAT policies and procedures. These protocols are now being applied to third-party manufacturers of finished products supplied to the company. These audits are much more like quality assurance, i.e., checking that management systems and procedures are in place, rather than quality control which is a more direct day-to-day responsibility.

Preparation of Statement and Internal Reports

For the first time, in 1995 The Body Shop decided to draft a full statement of its animal protection performance, particularly with respect to the AAT policy and procedures. The Animal Protection Statement is similar to the simultaneously prepared Environmental and Social Statements in that it provides the background to policies as well as information on how they are followed and statistics on supplier declarations and ratings based on in-depth research. In common with the other statements, there is a level of external verification (with respect to conformance with ISO 9002) and there is a commitment to strategic objectives and continuous improvement. The Animal Protection Statement also includes information on The Body Shop's position with respect to endangered animal species and the use of animal by-products in cosmetics. These issues are included for completeness but are not subject to any form of external validation.

Agreement of Objectives

Animal protection policies are somewhat less complex than environmental and social policies. They also involve fewer departments in their execution and enforcement. So the opportunity for target setting and the development of new activities is limited.

Nevertheless, as with other ethical audits, the principle of continual improvement is a vital component of the process. It is, therefore, included via an objective setting process involving relevant departments, and signed off by senior management.

External Assessment and Certification

The assessment of conformance with ISO 9002 was performed by SGS Yarsley ICS, an internationally recognized quality systems assessment group.

Publication of Statement

The first Animal Protection Statement was scheduled for publication alongside The Body Shop's annual Environmental Statement and the company's first Social Statement. Because of its novelty, steps have been taken to ensure staff familiarity with the process before publication. This was especially important for shop-based staff who are the people most likely to get questions on the subject of animal welfare and the cosmetics industry.

Seven Dos and Don'ts of Animal Protection Auditing and Disclosure

Do talk to animal welfare groups and pub-organizations who are campaigning against animal testing. They are happy to encourage and provide advice to companies who are serious about ending animal tests.

Do consider other animal welfare issues of relevance to your business. There is no point having a strong ani-position on animal testing if this is undermined by other animal welfare impacts.

Do ensure that departments, managers, and staff at every level understand the position of the company and are able to explain it to customers and other stakeholders.

Do set up appropriate monitoring and quality control procedures in relevant technical and purchasing departments. Make sure your audit system takes an independent view and reports directly to a main board director.

Do use respected assessors for verifying conformance with standards.

Don't make product claims or public statements without a clear understanding of the issues and means of delivering on your commitments.

Don't forget there are clear differences between conventional and organic farming with respect to animal welfare issues and clear distinctions between the vegetarian and vegan positions on animal welfare.

Don't forget to get staff mobilized into campaigns for alternatives to animal tests, and alternatives to cruelly harvested by-products. No one wants to hurt animals needlessly.

Don't forget to train your staff on your policies and procedures and check that operational staff are aware of any updates to policies and procedures.

Don't be afraid to ask animal welfare groups and consultants close to the animal welfare movements for advice on standards, systems, and certification.

Do adopt a formal standard and commit to it publicly.

Don't delay—especially if you are already making product claims on animal welfare.

Do report: formally and informally, publicly and internally. Stakeholder understanding is crucial to progress, as are targets and objectives for the future.

Don't be afraid of including both good and bad aspects of social performance; better that you draw attention to your faults than your critics do.

References

1. The Body Shop International, *The Body Shop Annual Report and Accounts*, Littlehampton, 1994.

2. Council for the European Communities, "Proposal for a Council Regulation (EEC) Allowing Voluntary Participating by Companies in the Industrial Sector in a Community Eco-Audit Scheme" [Com(91)459], *Journal of the European Communities*, vol. C76, 1992, pp. 1–13.

3. D. Wheeler, "Memorandum by The Body Shop International," *A Community Eco-audit scheme. 12th Report of the Select Committee on the European Communities*, House of Lords Paper 42, HMSO, London, 1992, pp. 58–59.

4. The Body Shop International, *The Green Book 1,2,3*, 1992–1994.

5. D. Wheeler, "Auditing for Sustainability: Philosophy and Practice of The Body Shop International," in L. L. Harrison (ed.), *The McGraw-Hill Environmental Auditing Handbook: A Guide to Corporate and Environmental Risk Management*. McGraw-Hill, New York, 1996.

6. DTTI, Sustainability and IISD, *Coming Clean: Corporate Environmental Reporting*, DTTI, London, 1993.

7. United Nations, *Agenda 21*, UN, Geneva, 1992.

8. S. Schmidheiny, with the Business Council for Sustainable Development, *Changing Course: A Global Perspective on Development and the Environment*, MIT Press, Cambridge, Mass., 1992.

9. Business Council for Sustainable Development, *Getting Eco-efficient: First Antwerp Eco-efficiency Workshop, Antwerp, November, 1993*, BCSD, Geneva, 1993.

10. P. Hawken, *The Ecology of Commerce: A Declaration of Sustainability*, Harper Collins, New York, 1993.

11. B. Taylor, C. Hutchinson, S. Pollack, and R. Tapper, *Environmental Management Handbook*, Pitman, London, 1994.

12. International Institute for Sustainable Development, *Business Strategy for Sustainable Development*, IISD, Winnipeg, 1992.

13. O. Williams, "Business Response to the Earth Summit," *ICC World Business and Trade Review,* 1994, pp. 123–124.

14. R. Gray and J. Bebbington, *Sustainable Development and Accounting: Incentives and Disincentives for the Adoption of Sustainability by Transnational Corporations,* University of Dundee, Dundee, 1994.

15. A. C. DeCrane, "Energy and Sustainable Growth," *ICC World Business and Trade Review,* 1994, pp. 127–128.

16. D. Wheeler, "The Future for Product Life Cycle Assessment," *Integrated Environmental Management,* vol. 20, 1993, pp. 15–19.

17. P. Shrivastava, "Ecocentric Management for a Crisis Society," *Proceedings of The Second Nordic Network Conference on Business and Environment, Oslo,* Norwegian School of Management, Oslo, 1994.

18. E. Callenbach, F. Capra, L. Goldman, R. Lutz, and S. Marbur, *Ecomanagement: The Elmwood Guide to Ecological Auditing and Sustainable Business,* Bernett-Koekler, San Francisco, 1993.

19. J. Carlopio, "Holism: A Philosophy of Organizational Leadership for the Future," *Leadership Quarterly,* vol. 5, issue 3/4, 1994, pp. 297–307.

20. E. Mayo, "Social Accounting," *New Ground,* vol. 41 (winter), 1994.

21. S. Zadek, *Making Business More Socially Accountable,* 1994.

5

Enhancing the Value of Environmental Audit Reports

Jane E. Obbagy
Vice President, Arthur D. Little, Inc.
Cambridge, Massachusetts

Corporate officers are taking a much more active role today in ensuring the success of their companies' environmental, health, and safety (EHS) programs. The reasons for this vary, but frequently stem from:

- Intense public scrutiny and media attention directed toward corporate management of EHS matters
- The threat of significant legal and financial liability and resulting loss of corporate reputation from a major EHS crisis
- A realization that environmental stewardship can contribute to competitive success and is increasingly a business strategy and planning issue

In addition, both the government and the public are pressing corporate officers to demonstrate that their companies are managing their EHS obligations effectively. In particular, companies are being pushed to provide for more disclosure of their EHS performance. Not surprisingly, senior executives are in turn asking their EHS staffs to provide them with assurance that:

- No substantive compliance violations exist
- Current operations do not pose any potentially serious threats to human health and safety or the environment
- Systems are in place and functioning to manage EHS compliance obligations and risks appropriately both now and in the future

The environmental audit—particularly through the written audit report—has taken on an expanded role as an important vehicle for providing assurance and informing corporate officers about the company's EHS progress or lack thereof. Audit reports are not the only source of information regarding the company's EHS performance. However, they are often among the most visible and, as such, provide EHS professionals with an opportunity for sharing valuable information with top management.

The Role of the Audit Report

For more than a decade, environmental auditing has been a valuable tool for providing EHS assurance within a corporation. As a result, choices about how to make audit report findings and their implications readily understood by top management take on a new importance; for example, what report format to use and what type and level of detail of information to include.

Virtually all environmental audits involve gathering information, analyzing facts, making judgments about the status of the facility, and reporting the results to the appropriate levels of management.

Information reported often includes an assessment of the strengths and weaknesses of EHS programs in place, as well as information regarding EHS compliance and other risk management issues requiring attention. These facility-specific findings can have value to management in and of themselves as an indication of how particular locations are carrying out their EHS responsibilities. When audit programs are staffed in a way that enables auditors to visit a number of facilities in the course of carrying out their duties, the auditors are in a unique position to provide an added perspective on how individual facilities can compare with each other. They can help provide a barometer of how EHS matters are being managed on a day-to-day basis across the corporation. Even an observation about a lot of little things wrong in facilities can be useful to management in suggesting that there is not enough attention being given to overall management of these EHS issues.

Thus, the environmental audit and its report present an opportunity to inform management more broadly about the company's EHS performance.

Presenting Audit Results

One principle for conducting effective environmental audits is that the audit report be prepared with an "appropriate" form and content. At a minimum, audit reports should present the purpose, scope, and results of the audit. However, audit reports that are highly valued within a corporation, relative to their role in providing assurance, go beyond the basics. They provide the right level and mix of information and contextual background to ensure that the managers receiving the report understand the implications of what is being reported. Moreover, these reports are written to draw the reader quickly into the importance of what is being discussed, and to identify the causes of underlying deficiencies in order to provide an effective trigger for implementing corrective actions, as appropriate.

Report Format

As a result of the growing role of audits in the mix of corporate EHS management tools, together with the increasingly direct involvement of management in EHS issues, over the last decade we have seen an evolution in how companies report their environmental audit results. Although there are a variety of report formats used today, a look at this evolution provides a useful benchmark for evaluating whether or not your company's audit report is consistent with the needs of management and the culture of your organization.

In the early days of audit program implementation, many companies were concerned that the written word might "come back to haunt" them. Thus, textual discussion in these audit reports—particularly interpretive information—was kept to a minimum. Audit teams generally produced *audit memos* that provided a basic message that "we visited the site and here are our general observations."

As companies became more confident both in the sophistication of their EHS systems and of the value of their audit programs as a management tool, the audit teams used the reports to provide a message that "we visited the site and here are the deficiencies noted." This *exception report* listed departures or exceptions from governmental requirements and company standards and provided observations related to the management of EHS programs reviewed during the audit.

With the increasing sophistication of EHS systems and audit programs, companies with procedures and checks and balances in place continued to list the exceptions and observations in audit reports. They also often included a generic statement about areas reviewed during the audit. The basic message of this *generic opinion report* was "we reviewed site programs and practices following our standard audit guidelines and procedures, and we believe the site is in general compliance except as noted in this report." This report implied a level of comfort to management, but because it always seemed to provide the same information, it continued to beg the question of just what the results meant for the company as a whole.

Taking the next step, today some companies are asking their audit teams to provide a "true opinion" of their overall analysis of the management systems in place and the level of compliance achieved with respect to a spectrum of relative performance, while also including a list of exceptions and observations ordered by their relative significance. The *true opinion report*, which is based on the professional judgment of the audit team, is gaining popularity among top management who count on audit reports for two main purposes:

- To help them understand relative overall EHS performance and the existence of specific problems

- To assist them in understanding the significance of the audit results so as to focus resources on areas where improvement is most needed

A characteristic of the true opinion approach is a preestablished range of opinions from which the team can choose with respect to the facility's overall performance in light of the available audit evidence. What contributes to the success of this report format is not the number of opinions used—four, five, or six—but the presence of a range of opinions that clearly reflects different levels of performance being achieved at a given facility. The opinions from which to choose could range, for example, from a relatively clean bill of health, such as "meets governmental and internal requirements," to a much more problematic reading, such as "requires substantial improvement to meet governmental and internal requirements." The team chooses the appropriate opinion and provides the EHS management analysis and audit evidence in the report to support that opinion.

Kind of Information

Consistent with the choice of a report format, the best audit reports provide readers with clear and succinct information and the right level of technical detail presented in an appropriate managerial context. We believe the following issues relative to report style are particularly important to consider:

- *Straightforward, succinct information*—Effective reports for management are best written to avoid any ambiguity. The clearer the information, the less judgment required by the reader to decide what is or is not an issue that requires attention. Moreover, reports should avoid terminology that is unlikely to be familiar to the addressee. For example, most senior managers are not conversant in regulatory jargon that is prevalent in environmental circles.

- *Appropriate level of technical information*—Taking into consideration the technical knowledge of the relevant management audience, the information provided should not only be clear but should also include explanatory and helpful technical background information as needed.

- *Contextual information*—Again, depending on the background and general knowledge of the audience with respect to EHS issues, it may be necessary to describe clearly the significance of particular points in order to alleviate any potential misunderstanding. Information should be presented so that the reader understands that a particular problem, for example, is a departure from a long-standing requirement. The report should also indicate to the reader whether or not the departure is from a governmental standard, company standard, or generally accepted good management practice.

Fine-tuning for Greater Value

Companies that are particularly successful in making the audit report valuable to management understand both the changing role of the audit report in providing assurance and also the characteristics of an audit that complement this role. Some helpful rules of thumb to consider include:

- *Find out just what top management wants the audit report to accomplish;* the answer may trigger a complete rethinking of your audit program and audit report approach. Including comments on the effectiveness of EHS programs is part and parcel of a *rigorous and structured approach to auditing.* Consequently, the standards established for the audit and staff chosen to participate as audit team members become more critical—consistency in audit team makeup from audit to audit is important.

- In order to meet the needs of senior management effectively, a *clear writing style* is important in an audit report. Companies should examine the resources typically devoted to the writing of audit reports—time allotted and the skill level of the individual actually doing the writing—to make sure they are sufficient to fulfill the expectations of the management audience to whom the reports are being sent.

Increasingly, it is senior executives who require assurance that their companies' environmental, health, and safety obligations are being met. The written audit report—used properly—can be the single most valuable tool for providing this information. A good place to start is to find out what management wants, and then to examine whether your audit approach and audit reports are sensitive to and complement those needs.

6

Auditing the Software Used in Environmental, Health, and Safety Programs

Norman P. Moreau

Management Analysis Company
Washington, D.C.

Computers have penetrated nearly every facet of an organization's operation. We rely on computers to design and run facilities, monitor process controls, manage maintenance, track environmental impacts, and provide evidence of regulatory compliance, just to name a few functions. If computers, and the software used to operate them, play such an important role, why is so little attention paid to the auditing of software? The short answer is that software remains a mystery to many people, auditors included. Mystery leads to fear, and fear leads to avoidance. Consequently, few audits examine the development, operation, and maintenance of software. When deciding whether or not to audit software, consider what the impact of the following real examples of software-related mistakes would have on your facility:

- Piping software contains error in calculating minimum, maximum, and average transit times of fluids.

- Vendor loads an uncontrolled version of software in an accident mitigating circuit resulting in an incorrect time delay.

- Missed calibration on lower explosive limit monitors caused by erroneous input into tracking database.

Software Definitions, Applications, and Standards

Software Definitions

When discussing any discipline it is important to have an agreed upon set of terms and definitions. The Institute of Electrical and Electronics Engineers (IEEE) publishes a compendium of software standards called the *IEEE Standards Collection: Software Engineering.* The collection is an excellent desk reference used by many software professionals. Within the collection is a general glossary of software engineering terminology (IEEE Std. 610.12). The standard defines software as composed of computer programs, procedures and possibly associated documentation, and data pertaining to the operation of a computer system.

For our discussion, the definition of software contains another term that requires defining, i.e., computer program. A computer program is a combination of computer instructions and data definitions that enable computer hardware to perform computational or control functions. Software is the broad scope definition that contains not only the computer program (commonly referred to as the program or code) and data but also associated documents. The reason for presenting these definitions is that the auditor of software needs to have a basic understanding of the terminology used to communicate in this discipline. Communication skill is a key attribute of an auditor.

Application of Software

Many types of software can be found in an organization. Software types include spreadsheets for performing simple calculations, computerized maintenance management systems, distribution control systems for plant operations, and accounting and financial products. Is the EHS auditor concerned with all software within an organization? Absolutely not.

Table 6-1. Classification Scheme

Software applications generally include:
Computers used for the control of plant equipment or components, or providing direct indication of plant status
Computers and computer programs used for modeling the operation of plant equipment for design activities, scientific analyses, or calculations
Computers and computer programs used to demonstrate regulatory compliance or commitment
Information systems used for the purpose of administrative control

Software by itself does not create EHS hazards. It is the end use or application of the software that creates the hazard. Consider software developed for use in a continuous stack emission monitoring device. If the software contains a defect and an incorrect logic path is taken that leads to a release of hazardous material, the health liabilities, regulatory fine, and associated public discontent could be significant. The same defect in a nearby device used to monitor stack temperatures may only result in a minor operational inefficiency. Because the defect in the first device may result in a significant hazard, the application of this type of software warrants a classification greater than the second device. The auditor should expect to see the controls applied to the first device more stringent than the second device.

Organizations typically have procedures that describe how software they use is identified and how the software is classified based on its application. The results of the classification process can guide the auditor in selecting software products that will be subject to audit. Table 6-1 shows a typical classification scheme.

Software Standards

G. H. Mealy stated in 1969 that:

> Standards are like morals; they are rules adopted by some segment of society, within some context of social behavior, to regulate activity—everything else being equal. But, they are not the same in all contexts, nor should they be regarded as hard and fast within any given context. We erect them so that people will usually behave in a predictable way. In a [computer] system development context, however, another valid purpose for a standard is to subject a proposed deviation from the standard to public scrutiny in order to decide whether the deviation is justifiable within the total system context.

A standard is a rule or basis for comparison that is used to assess the size, content, value, or quality of an object or activity. In software, two kinds of standards are used to define the way software is developed and maintained. One kind describes the nature of the object (i.e., software) to be produced, while the other defines the way work (i.e., development of the software) is to be performed. Standards also exist for programming languages, coding and conventions, commenting, error reporting, and so forth.

A standard is appropriate when no further judgment is needed. Standardization makes sense when items are arbitrary and must be done uniformly or when there is one clearly best alternative. The definition of coding or naming conventions, the selection of a programming language, and the selection of a design method are good candidates for standardization.

There are many cases when standards are totally inappropriate. Typically these are cases involving technical judgment. One such example is the specification limits on module size. The evidence is overwhelming that large modules are likely to be complex and hard to maintain. A standard that states that no module can exceed 300 lines of code may not be a wise practice. There is conflicting evidence on the effect of module size on the quality or cost of programs, while there is clear evidence that too many small modules cause performance problems. There is no question that the computer program should be constructed in modules, but this is too complex a question for arbitrary rules. In questionable cases like this, a guideline should be used or the standard should have practical escape provisions. The auditor must recognize that just as good engineering judgment is an acceptable practice in plant design, technical judgment is an acceptable practice in software development. In both cases the auditor can and should ask for the basis of that judgment.

Procedures are closely related to standards and guidelines, and are often developed directly from them. Table 6-2 shows several instances of a standard or guideline, and a procedure for essentially the same topic. The distinction between the two is that the standard describes content, etc., while a procedure describes how the work is actually to be done, by whom, when, and what is done with the results.

Auditing Software

Auditing software is different from auditing hardware. It requires insights into the unique aspects of the software development process and the particulars of software development, operations, and mainte-

Table 6-2. Representative Software Standards, Guidelines, and Procedures

General	
ISO 9000-3 "Guidelines for the Application of ISO 9001 to the Development, Supply and Maintenance of Software"	Software Quality Assurance Manual
Software Configuration Management (IEEE 828 and 1042)	Software Change Control
Software Verification and Validation (IEEE 1012 and 1059)	Software Verification and Validation Plan
Software Reviews and Audits (IEEE 1028)	Steps for Auditing the Testing Process
Software Problem Reporting and Corrective Action (IEEE 1044)	Change Control Authority
Software Acquisition (IEEE)	Software Procurement Specifications

Development	
Software Quality Assurance Plan (IEEE 730)	Preparing Software Quality Assurance Plan
Software Metrics (IEEE 1045 and 1061)	Establishing Software Project Metrics
Software Requirements Specification (IEEE 830)	Software Requirements Specification Review
Software Design Description (IEEE 1016)	Software Design Review
Software Coding Conventions (ANSI Std.)	Programming Language Manual
Software Testing (IEEE 829 and 1008)	Unit Level Testing

Maintenance	
Software Maintenance (IEEE 1219)	Software Modification Request
Software User Documentation (IEEE 1063)	Verification of User Documentation

nance. To add value to the software audit, the auditor must be sensitive to these unique aspects. The software development process is the set of tools, methods, and practices used to produce a software product. The auditor must recognize which management system principles are directly applicable, which may be treated by analogy, and which are not directly applicable at all.

Reviews versus Audits

Reviews and audits are events where outsiders analyze and critique either the evolving software product or the software development process or both. In software, particularly during development, it is important to distinguish between reviews and audits. With reviews, it is simple. Management decides what reviews are needed during the planning process (e.g., requirements review, design review, code review, validation review, etc.) and puts onto the project schedule control points where the reviews need to be accomplished.

With audits, the answer is case specific. During development, audits by the user will vary depending on the application and complexity of the computer program. Through its normal project monitoring process, management of the project or user organization may spot an anomaly which deserves further study. In either case, an audit team is established and presented with an audit purpose and scope, and given resources to complete its assignment. When its findings are presented, the audit team is disbanded. Table 6-3 shows the similarities and differences between reviews and audits.

Types of Software Audits

As with other types of audits there are basically two types of software audits, the project audit and the process audit. The project audit is usually performed during the development of software. Project audits occur because a perceived problem exists or because the user has speci-

Table 6-3. Reviews versus Audits

Reviews	Audits
Generally for a review, it is vital to have:	Generally for an audit, it is vital to have:
Qualified and available team members	Qualified and available team members
Resources to enable the team to perform	Resources to enable the team to perform
A predefined agenda	An audit purpose and scope
Review criteria	A commitment from the auditee that they will not obstruct the audit team's search
A clear picture of the goals of the review	The authority to enable the use of the findings of the audit

fied the need for one in a contractual document. A perceived problem may exist for a project that is deeply over budget and way behind schedule. A project audit is one method that provides an objective view that may lead to a solution.

Depending on the size of the organization, software development projects are often provided by a supplier. The supplier may be under contract or may be the organization's Information Systems (IS) group. In either case the term *supplier* will be used synonymously. The project audit is typically performed with the supplier's resources, but this should not restrict the user organization from conducting periodic project audits. Regardless of the organization performing the work, every project should be audited to ensure honest and competent performance. Without some kind of independent review, projects rarely perform with the discipline required or specified. This is not because people are dishonest, but because the pressures of software development are so enormous and the apparent efficiencies of most shortcuts are so seductive. Thus a system of checks and balances that includes project management and user audits is needed to reduce the temptation of shortcuts.

Two other project audits that are often performed by the software supplier are the functional audit and the physical audit. Both audits are performed by the developer and occur prior to the delivery of the software. The objective of the functional audit is to provide an independent evaluation of software products, verifying that its configuration items' actual functionality and performance are consistent with the requirements specifications. The objective of the physical audit, which is also an independent evaluation of a software product configuration item, is to confirm that components in the as-built version maps to the specifications. These types of audits are easily adapted to software maintenance activities where risk of the software application is great or the modification was significant.

The other type of audit is the process audit. EHS audits are typically process audits with the objectives of verifying compliance, assessing the EHS management system, identifying EHS hazards, and recommending program improvements. The objectives of process audits of software are no different. The software audit seeks to verify compliance with specifications, standards, and procedures; assess the software management system as it impacts the EHS application; identify EHS hazards that may result from the application of the software; and recommend program improvements that result in software that meets the objectives of the EHS program.

Software Auditors

Software Sensitivity

Auditors of software should possess software sensitivity. The ideal auditor of software is an information technology (IT) subject matter expert with the education and experience to audit the software development and maintenance process. Unfortunately, individuals with these skills are either consultants or not interested in becoming auditors. In many organizations concerned about the impact of software, management directs the use of these subject matter experts as part-time auditors. Programs such as these have been successful and proven useful to both the subject matter expert and other audit team members.

For those who are not subject matter experts, there is a need to create awareness and sensitivity to fundamental software concerns. Software sensitivity issues include the following:

1. Impact that software has on the EHS program

2. Issues of configuration management, including identification, status accounting, access control, and change control

3. A comprehensive understanding of the software development life cycle

4. How software fits into the EHS management system for any plant component, product, or service

Qualifications

To define the qualifications for a software auditor, an organization could examine the qualifications requirements of a TickIT auditor used in the ISO 9000 registration process. In the late 1980s, the U.K. Department of Trade and Industry commissioned studies that highlighted both the crucial role of IT and the need to address quality problems within IT. Out of these efforts came a project to apply ISO 9000–based assessment and registration to the particular nature of IT. The project was dubbed TickIT, a visual/verbal play on the "tick" (check mark) for "IT" being an admission ticket to higher quality.

The TickIT scheme established specific experience and training requirements for quality system auditors to provide more software-knowledgeable individuals. If a registrar wants to include IT within its scope, it can only use auditors qualified specifically within the TickIT scheme. Individuals applying for certification as software quality system auditors are required to possess qualifications such as those shown in Table 6-4.

Table 6-4. TickIT Auditor Qualifications

Academic and training	Experience
Academic degree or additional supplementary workplace experience if less than bachelor's (two more years for associate degree and five more years for secondary diploma)	At least five years of relevant software workplace experience, with at least two of the past four years in quality activities ("relevant" means addressing full software life cycle, software quality management systems, software auditing or detailed knowledge of one area of software expertise)
Completion of a week-long approved Software Quality System Auditor training course	Experience in quality system audits against ISO 9001, ISO 9002, or equivalent national, industry, or company standards

Software Auditor Training

The TickIT-approved software quality system auditor training course provides an excellent base for qualifying software auditors. An organization with qualified audit personnel that is contemplating using increasingly complex systems with new emerging technologies should start making plans for the training and development of their personnel.

Since most organizations use their auditors in pressing and current audit projects, this means that very seldom is any staff time set aside for the development of the new skills that will be required to audit new and complex technologies. When the technologies arrive, the audit manager will be faced with training their staff on an after-the-fact basis. Most of all, they would be unable to provide timely and necessary strategic audit input while the technology is being selected, which should increasingly involve the EHS auditing group.

Steps to consider in the training of future audit professionals include:

1. Identify the potential technology planned for adoption. This can be accomplished by keeping in close contact with engineering and IS/IT groups responsible for technology introduction and planning. Someone in EHS auditing should become a participant in the key presentation sessions of these groups.

2. Entrust someone in the audit group to determine the audit implications of the new or changed technology. Implications include con-

trols likely to be needed, new or modified audit approaches, and the training program for personnel in the audit group.

3. Develop the specifics of training and development down to types of courses, sources of training, and individuals who should take the training.

Audit Team

The audit team should consist of team members who are software sensitive and have received suitable levels of training in the technology to be audited. The team should be represented by at least one member familiar with the organization being audited. For example, an audit of the software associated with a plant's distribution control system would include a member from operations and computer support services. It is always to the team's advantage to have a member from the engineering department. The engineering groups' insight can be extremely helpful in understanding the application of software. For large projects, the supplier audit team should be represented by someone involved in writing the requirements specification. Since the specification drives the direction of development, this member can be an invaluable resource for clarifying specification intent and resolving problems.

The audit team leader should also be software sensitive and have received the same training in the technology being audited as the audit team members. The training requirements are less critical for the audit team leader, but participation should be encouraged. A cautionary note in team selection: For project audits where management has identified a problem or has concerns, team members should not include staff members directly involved in the project. The staff member may be a contributing cause to the problem and is now in a compromising situation. The best approach here may be to use a consultant.

Life Cycle Activities

Software has a life cycle that begins with a concept and ends when the software is retired. Development, operations, and maintenance of software can occur at the facility, through the corporate IS/IT departments, or though a contract with a supplier. In each case, the life cycle activities are the same. The life cycle paradigm calls for a planned and systematic approach to software development that begins at the concept or system level and progresses through requirements, design, implementation, testing, installation and checkout, operations and maintenance, and retirement. The activities the software goes through are depicted in the classic life cycle called the *waterfall life cycle model.* The waterfall model,

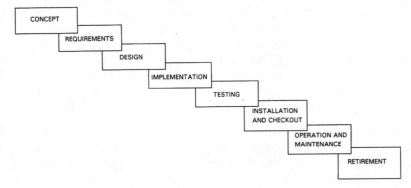

Figure 6-1. Representative software life cycle activities.

shown in Fig. 6-1, has received criticism for various reasons. The problems encountered when the waterfall model is applied include:

1. Real projects rarely follow the sequential flow that the model proposes. Iteration of activities always occurs and this creates problems in the application of the paradigm.
2. It is often difficult for the customer to state all requirements explicitly. The waterfall model demands that requirements be explicitly stated, and has difficulty accommodating the natural uncertainty that exists at the beginning of many projects.
3. Customers are not patient and want working versions as soon as possible. Strict adherence to the waterfall model restricts the delivery of a working version of the program(s) until late in the project. A major error or mistake left undetected until the working program is reviewed can be disastrous and costly.

The EHS auditor needs to recognize that other life cycle models such as prototyping and spiral have emerged and are being used with a great deal of success. However, it is also important that the auditor verify that the organization has established a life cycle approach and that appropriate controls are defined. While the waterfall model does have weaknesses, it is significantly better than a haphazard approach to software development. For the purpose of this discussion, the basic waterfall model will be used.

The sections that follow introduce common terms, activities within the life cycle, and activity attributes. Each activity has recommendations that the EHS audit manager and software auditor should consider in the auditing of software. The auditor should be aware that the activities and attributes presented reflect preferred conditions and should be viewed as a guide to good practices in the development, operations, and maintenance of software. When auditing software it is important that the auditor focus on the agreed upon project requirements and implementation of organizational policies and procedures.

Concept

The concept activities normally result in high-level documents that focus on the system as a whole, one part being software. The concept activity should involve the user organization. The EHS audit team uses the documents generated from this activity to understand the application of the software and its importance to the EHS program.

Requirements

This is the first activity of the software development cycle. It is here that the user's problem is stated, analyzed, understood, and translated into a problem definition. The result of the requirements activity is a requirements specification for the software. This requirements specification is often referred to as a software requirement specification. The audit team should have a thorough understanding of the content of this document. All functions that the software is to perform are found in the requirements specification. A good requirements specification should be correct, unambiguous, complete, consistent, verifiable, modifiable, and traceable.

Any complex computer system is the concern of a number of organizations. The organizations involved must reach the same understanding of the meaning of the product. Consider the following example which is based on an actual incident from the early days of computerized process controls. The requirements specification for the plant read, in part: "in case of an alarm, hold the variables constant and call the system operator." The relevant portions of the system are shown in Fig. 6-2. The computer controls valves that admit catalyst into the reactor vessel and cooling water into the condenser. The computer can receive alarm signals from various components of the system, including the low oil level alarm. The low oil level alarm is shown as coming from one of the gear boxes in the system.

When the system was first put into operation, it failed with the following scenario:

1. The reactor was charged.
2. The catalyst valve was opened to start the reaction.

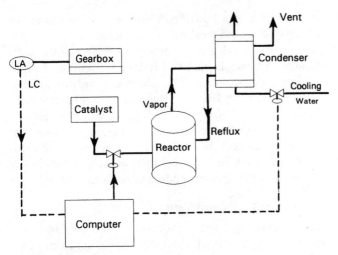

Figure 6-2. Diagram of chemical plant. (*Source:* "Human Problems with Computer Control," *Hazard Prevention,* March–April 1993, pp. 24–26.

3. The cooling valve was closed to allow the reaction to reach operating temperature.

4. A low oil alarm was sensed (this later turned out to be a spurious condition).

5. The values were held as indicated, the catalyst valve open and the cooling water valve closed, while the operator was notified of the (spurious) low oil condition.

6. The reaction proceeded unchecked.

7. The reactor overheated and vented its contents.

The root cause of the mishap was the failure of the chemical engineer who designed the system and the programmer who implemented it to reach a common understanding of the term "variable." To the engineer, the variables of the system were the temperature and pressure in the reaction vessel. To the programmer the variables were the valve positions under the control of the program. The two views are linked by the control laws of the reactor system and the chemistry of the particular reaction under control, but these were not part of the requirements specification.

While this story appears to exhibit an almost unbelievable lack of understanding on the part of both the engineer and the programmer, many developers of requirements specifications seem determined to create language that can easily lead to similar situations.

Auditing the requirements activity is an important aspect of the software audit. Without adequate requirements it will be difficult to design, code, and test the computer program. The software auditor should focus attention on the depth and breadth of the requirements specification and be prepared to ask the unforgiving, obvious question. Challenge the specification writers to be concise and clear. Propose scenarios that cause them to answer the what-if-you-were-not-here-to-answer-this-question question—can the specification stand on its own? Closely examine the review of the requirements specification:

1. Was the review made up of only software specialists?

2. Did the team include a real user, that is, someone from the organization that will have to live with the product, e.g., someone from operations or maintenance?

3. Can requirements be tested, or are terms such as "easy to view," "respond quickly," or "within a reasonable time" used? Remember, the clearer the requirements specification is, the easier it will be to provide a traceable audit trail should the software be suspect during an EHS mishap.

Design

This is the second activity of the development cycle. Here the "what" of requirements are translated into the "how" of a solution. The design process translates the requirements specification into a representation of the software that can be assessed for quality before coding begins. Computer system specific decisions are made: What computer? What language? What modules? What sequence of functions? What data structures? The product of the design activity is a software design description.

The software design focuses on three distinct attributes of the computer program: data structure, software architecture, and procedural (module) detail. A good design should:

1. Exhibit a hierarchical organization that makes intelligent use of control among the elements of software.

2. Be modular; software should be logically partitioned into elements that perform specific functions and subfunctions.

3. Contain a distinct and separable representation of data and procedure (module).

4. Lead to modules (e.g., subroutines or procedures) that exhibit independent functional characteristics.

5. Be derived using a repeatable method that is driven by information traceable to the requirements specification.

These attributes are normally verified through the review process. The software auditor is not expected to repeat the review team's effort; however, the auditor should verify the review process and that the review included the above attributes.

Implementation

The third activity of the development cycle is implementation. The design is translated or coded into a computer-readable, computer-processable solution. Here as the computer program takes shape the "how" of the design becomes a problem solution. Attributes of good computer programs will vary depending on the programming language used. Three common attributes include portability, maintainability, and testability.

Here is where a subject matter expert comes in handy. The "ility" attributes described are often difficult to evaluate. They tend to be somewhat subjective and only a set of expert eyes will recognize difficulties. One of the activities performed to maintain the computer program is commenting, that is, the insertion of descriptive language within the computer code. Commenting within the computer program provides the software maintainer with specific information about a portion of the computer program. Poorly commented computer programs make the computer program difficult to maintain in the future. Unless specified within a guideline or procedure, the auditor or subject matter expert must rely on their experience to defend a finding on insufficient commenting.

It is during the implementation activity that many of the organization's standards, guidelines, and procedures can be verified. Such documents include naming of conventions, limits on code complexity, code element traceability to requirements, and development of the user's documentation.

Testing

This is the fourth activity of the development cycle. Once code has been generated, computer program testing begins. Here the computer program is run to see if it meets the requirements. The testing process focuses on the logical internals of the software. Software testing is performed to detect programming and design errors, question unrealistic requirements, and put the final touches on the soon-to-be-usable computer program. Testing should be carried out according to test plans, cases, and procedures. Results of tests should be recorded and evaluated according to the test plan.

Testing begins at the module level and spirals outward as the software is integrated into the computer system. Unit testing makes up the inner level of the testing spiral. Unit testing normally occurs at the module level and is performed during the implementation activity. As modules are combined, integration testing begins. The focus of integration testing is on design and the construction of the software architecture. After integration testing is complete, the spiral continues its outward track and validation testing begins. Here the software is tested to assure that all functional and performance requirements are met. Depending on the complexity of the project, validation may be the first testing that is formally documented. Up to this point, software defects are treated as problems and tracked to resolution by the programmer. Defects identified during formal testing are tracked to resolution through the configuration control process. It is important to agree upon the level of controls applied during testing. A pre-award supplier audit or survey is an excellent opportunity to work out these details.

When the integrated systems are combined, system testing is performed. Validation and system testing is often carried out by a group independent of the developer. System testing may also be referred to as acceptance testing. When performed at the developer's facility it is known as the *factory acceptance test*. Testing performed in the actual operating environment is known as *site acceptance testing*. It is during the system test that the supplier will initiate the functional and physical audit.

Testing is a very busy time in a project. Deadlines are fast approaching and everyone is concerned about deliverables. Standards and procedures are often geared to meeting the schedule. The EHS audit manager should consider assembling an audit team during the validation and system testing effort. The developer should have prepared a software verification and validation plan describing the goals and objectives of the validation effort. In order not to impede the testing process, the audit team needs to have a clear understanding of the verification and validation plan, and the test plan and procedures before the audit begins.

Many of the activities described in the testing of software during development are equally suitable during maintenance. An audit of testing activities during operations and maintenance is an appropriate method of verifying compliance with facility standards and procedures.

Installation and Checkout

This is the fifth and final activity of the development cycle and is viewed as the turnover or transfer activity. It involves checking the deliverables against the configuration item list and building or installing an executable system in the operating environment. Once installed the site acceptance test can begin. The site acceptance test demonstrates the capability of the software in the operational environment and validates that it meets the user's requirements as stated in the requirements specification. Results of the site acceptance test must be recorded and reviewed.

As installation approaches, the user organization becomes eager to receive working copies of their new or modified product. It is common for the user to receive portions of the computer system before the installation and checkout activity begins. Computer systems are not normally subject to the typical receiving inspection process. This results in computer components, computer programs, and documentation being tracked by the user but not subject to the formal controls of the nonconformance system. The EHS audit manager may want to add the installation and checkout activities to the audit performed during validation and system testing. The audit team should ask if deliverables received before the formal transfer activity began will be subject to the same level of testing as other computer components. The process to control these deliverables should also be evaluated. In the case where fragmented acceptance is planned, the audit team should verify that the final computer system accepted by the user has been adequately controlled regardless of when the transfer occurred.

In a real-time operating environment, computer systems may not operate as they did during the factory acceptance test. During site acceptance testing, the developer may have the ability to make on-the-spot corrections; however, without implementing the facility's or developer's change control procedures, latent errors may be introduced that may prove catastrophic later. Witnessing the installation and checkout activity goes a long way in preventing this sort of temptation.

During the checkout activity the user's documentation should be validated. Validation is best performed by the actual operator of the computer system.

Operation and Maintenance

When development is finished, the operations and maintenance activities of the life cycle begins. The operational process involves the user operating the system. The operator is expected to use the system in its intended environment and according to the user's documentation and operating instructions.

Another area to consider during the use of software is in-use testing. In-use testing is particularly important for process control software such as elements of a distribution control system and programmable logic controllers. This type of testing permits confirmation of acceptable performance of computer programs installed in operating systems. In-use testing should be conducted in a similar fashion as other testing using test plans, cases, and procedures. Plant surveillances are a typical method used to perform this type of testing. Many of the newer systems contain automatic self-check routines, minimizing the need to perform the periodic manual tests. The software auditor should include a random sampling of in-use testing documents and review selected equipment results as part of the facility's software audit.

Maintenance of software can account for more than 50 percent of all effort expended by a developer. Therefore it is imperative that the user has a program in place for modifying existing software. The maintenance activity is concerned with the resolution of software errors, faults, failures, problems, and computer program improvements. Depending on the type of change, the life cycle activities described in the preceding sections apply to software maintenance.

Software maintenance consists of four activities that are undertaken after a computer program is released for use. The first is corrective maintenance used to correct latent errors. The second is adaptive maintenance which is an activity that modifies software to properly interface with a changing operating environment. The third activity is perfective maintenance which occurs when new capabilities and enhancements are desired. The last activity is emergency maintenance which is a form of corrective maintenance that is unscheduled and is performed to keep a system operating.

Auditing a facility's software maintenance program is a comprehensive undertaking. The facility's computer support services group will be a key interface. A facility's change control authority may be another excellent source of information. The audit must include suppliers that provide both on-site and off-site maintenance services. A sample of the questions that need to be answered during a software maintenance audit include:

1. Who is responsible for the software maintenance program?
2. To what level are maintenance responsibilities delegated?
3. Is the software modification process being used and is it effective?
4. Who handles emergency maintenance, particularly on backshift?

5. Are emergency patches or fixes documented and properly processed within the specified time frame?

6. What controls the work performed by suppliers, both on and off site?

7. Is regression testing considered, performed, and documented?

8. Who and what determines when software has been modified too many times?

The list goes on. The point is that in planning such an audit allow sufficient time for the audit team to do research and prepare checklists. The lead auditor should be a senior staff member and be familiar with the organization and organizational interfaces.

Retirement

This is the final activity of the life cycle. The software is no longer needed or supported. The software is removed from all active systems and archived according to the facility's procedures. The auditor should verify how the facility identifies software users and verify that the software is no longer in use. Verification of archiving activities should not be limited to the documentation associated with the software, it should also include the computer program. Typical questions include: What are the access controls? What is the storage media? If needed, is the computer program retrievable and on what hardware platform can it be run today?

Support Activities

There is a great deal more to the life cycle of software than development, operations, and maintenance. The activities of the life cycle must be closely managed if the project is to be successful. This section describes the activities that parallel the life cycle activities. These activities include configuration management and verification.

Software Configuration Management

From the definition of software, we find that the output of the life cycle activities can be divided into three broad categories: (1) computer programs, (2) documents that describe the computer program, and (3) data structures. The items that compose all of the information produced as part of the life cycle process are collectively called the software configuration. As software progresses, the number of software configuration items grows rapidly. If each item only leads to another item, little confusion would result. Unfortunately, another variable is introduced: change. Change may occur at any time and for any reason. Software configuration management is a set of activities developed to manage change throughout the life cycle.

Software configuration management consists of four principal activities. The first activity is identification of the software configuration items. A variety of automated tools have been developed to aid in the identification activity. An identification process ensures that meaningful and consistent naming conventions for all items in the software configuration are applied. Information used to identify configuration items include document type, project or product identification, version number, release dates, etc. The auditor should verify that configuration items are identified consistent with the standards and procedures used by the facility.

Configuration items are usually stored in a software library. The software library is a repository for controlling configuration items. Adequate security and access control is essential for software libraries.

The most important activity of software configuration management is configuration or change control. The configuration management concept that helps control change without seriously impeding the change process is known as *baselining*. Before a configuration item becomes a baseline, change may be made quickly and informally. Once a baseline is established, a formal procedure must be applied to evaluate and verify the change. Agreement on when a configuration item becomes a part of the baseline is often difficult to reach. The developer prefers to baseline after the completion of all the testing, while the user organization prefers baselines to occur after agreement is reached on each activity of the life cycle, i.e., requirements, design, etc. The auditor must know what document describes the baseline requirements and when a configuration item becomes a part of the baseline.

Configuration control applies throughout the life cycle, i.e., during development, operations, and maintenance. The configuration control process begins when a change request is submitted and evaluated to assess the technical basis, potential side effects, overall impact on the system functions, and projected cost. The results of the evaluation are presented as a change request report that is used by the configuration control board or change control authority who makes a final decision on the status and priority of the change. To provide an adequate level of confidence in the configuration control process the auditor should evaluate the level of compliance with the prescribed process. One technique to verify compliance with the configuration control process is to compare an active configuration item with the most recent version in the software library. If the configuration does not compare, changes such as emergency patches have been introduced without proper controls.

Configuration status accounting is the third activity of software configuration management. Configuration status accounting involves the tracking and reporting of all configuration items and is conducted throughout the life cycle. Particularly in large projects or for control of many computer systems, configuration status accounting plays a vital role. Without an effective status reporting system the left hand not knowing what the right hand is doing syndrome will occur. The audit should verify that development or maintenance teams are apprised of important changes through regularly generated status reports and are working with current versions of the software configuration.

The fourth activity of software configuration management is the configuration audit. Configuration audits determine to what extent the actual configuration item reflects the required physical and functional characteristics. These audits are normally performed as part of the development or maintenance process. The EHS auditor should verify that proper auditing techniques were used and that problems identified are correctly handled through the corrective action process.

Verification

Verification refers to the set of activities that ensure that software correctly implements a specific function. A term often used simultaneously to verification is validation. Validation refers to a different set of activities that ensures that the software built is traceable to requirements. Simply stated, verification is about the reviews performed throughout the life cycle, and validation is about the testing performed at the end of the development process. One verifies or reviews the results of validation.

The level of verification depends to a great extent on the application and complexity of the computer program. Verification activities are performed as each activity in the life cycle nears completion. Examples of verification activities include reviews (walkthroughs, software inspection, and technical reviews), testing, and audits. The EHS group should play an active role in the verification process. This includes development and review of the verification and validation plan. The EHS audit manager needs to communicate that early involvement is not audit entrapment, but a method for problem solving early in the process when cost and schedule do not inhibit the production of quality software.

Auditing the verification process involves understanding the requirements of the software and the verification plan content. The auditor should verify the qualification of reviewers, the level of instruction provided reviewers, and the methods for documenting and resolving comments. Other considerations include determining what happens to unresolved comments and whether comments that result in changes to baselined configuration items are processed through change control.

Summary

Software will continue to play an important role in environmental, health, and safety programs. To minimize risk and enhance a facility's overall effectiveness and efficiency, the facility needs to consider the impacts of software. The EHS audit schedule should include audits of software based on the application and complexity of the computer program. EHS audit managers need to sensitize their auditors to the unique requirements of software and work to have the audit group actively involved in software development, operations, and maintenance activities. Finally, because auditing is different, the software auditors need to ensure that their auditing skills recognize the unique aspects of software.

Further Reading

The EHS auditing group should have a set of references to help the sensitization effort. The following is a small sample of the references available that may help the auditor of software.

Stephen B. Compton and Conner, Guy R., *Configuration Management for Software,* Van Nostrand Reinhold, New York, 1994.

Roger S. Glass, *Building Quality Software,* Prentice-Hall, Englewood Cliffs, N.J., 1992.

Watts S. Humphrey, *Managing the Software Process,* Addison-Wesley, Reading, Mass., 1990.

IEEE Standards Collection: Software Engineering, IEEE, New York, 1994.

Capers Jones, *Assessment and Control of Software Risks,* Yourdon Press, Englewood Cliffs, N. J., 1994.

Javier F. Kuong, *Computer Auditing, Security, & Internal Control Manual,* Prentice-Hall, Englewood Cliffs, N.J., 1987.

Robert W. Peach, *The ISO 9000 Handbook,* 2d ed., CEEM Information Services, Fairfax, Va., 1995.

Roger S. Pressman, *Software Engineering: A Practitioner's Approach,* 2d ed., McGraw-Hill, New York, 1987.

J. Sanders and Eugene Curran, *Software Quality,* Addison-Wesley, Reading, Mass., 1994.

TickIT is fully explained in the *Guide to Software Quality Management System Construction and Certification Using EN 29001.*

Index

About the Editor

Lee Harrison has been involved in environmental issues professionally since graduating from Northeastern University with a degree in civil engineering in 1969. He has written extensively on this subject as environment editor of *Chemical Week,* publisher of the *Environmental Audit Newsletter,* and editor of the first and second editions of the *Environmental, Health, and Safety Auditing Handbook.* Harrison currently is a consultant, advising corporate managers on crisis avoidance, preparation, and management.

Environmental Management

$39.95 U.

*Up-to-the-minute information on the latest standards
for environmental, health,and safety audits*

This essential update to McGraw-Hill's *Environmental, Health, and
Safety Auditing Handbook, Second Edition*, covers all of the regulatory
changes and new auditing standards implemented since the book was
published. Accessible and free of legal jargon, it is a must for anyone
responsible for compliance with federal, state, and local environmental
laws and corporate environmental standards and objectives.

Look inside for:

- The most recent regulatory information and auditing standards,
  including ISO 14000 and 14001

- New techniques and methods for conducting effective audits and
  re-engineering the EHS audit process

- Innovative approaches to measuring sustainability in business

- A guide to EHS auditing within a decentralized management structure

Learn how to plan and conduct effective environmental audits with
step-by-step guidance from leading corporations and consultants with
extensive experience in EHS auditing. This valuable reference work
guides you through the process, with sections on auditing and the environmental management system ... international approaches to EHS
auditing ... enhancing the value of EHS audit reports ... using computers
in EHS auditing, and even auditing the software used by EHS auditors
... and environmental audits and the law.

This information is detailed in language you can understand to help
you conduct EHS audits that are effective in measuring compliance,
evaluating management systems, and helping your organization
achieve its environmental, health, and safety goals—whether your
organization has a long-established EHS audit program or is just beginning
to develop one.

Cover Design: Richard Adelson

P/N 026921-1
PART OF 0-07-026921-1
ISBN 0-07-913651-6
          90000

9 780079 136510

McGraw-Hill